Zirconia - New Advances, Structure, Fabrication and Applications

Edited by Uday M. Basheer Al-Naib

Published in London, United Kingdom

Zirconia - New Advances, Structure, Fabrication and Applications
http://dx.doi.org/10.5772/intechopen.103991
Edited by Uday M. Basheer Al-Naib

Contributors
Ali Dahee Malallah, Nadia Hameed Hasan, Alexander Chee Hon Cheong, Sivakumar Sivanesan, Bincy Cyriac, Ghassan Albarghouti, Haneen Abdulkarim Sadiq Sadi, Hanumantha Murali Rao, Mamtha Kumaraswamy, Dhanu Thomas, Shivakumar Boraiah, Kuldeep Singh Rana, Sojood Mahmoud Farhan Darwish, Uday M. Basheer Al-Naib

First published in London, United Kingdom, 2023 by IntechOpen
IntechOpen is the global imprint of INTECHOPEN LIMITED, registered in England and Wales, registration number: 11086078, 5 Princes Gate Court, London, SW7 2QJ, United Kingdom

British Library Cataloguing-in-Publication Data
A catalogue record for this book is available from the British Library

Additional hard and PDF copies can be obtained from orders@intechopen.com

Zirconia - New Advances, Structure, Fabrication and Applications
Edited by Uday M. Basheer Al-Naib
p. cm.
Print ISBN 978-1-83768-897-5
Online ISBN 978-1-83768-898-2
eBook (PDF) ISBN 978-1-83768-899-9

For EU product safety concerns:
IN TECH d.o.o., Prolaz Marije Krucifikse Kozulić 3, 51000 Rijeka, Croatia,
info@intechopen.com or visit our website at intechopen.com.

Meet the editor

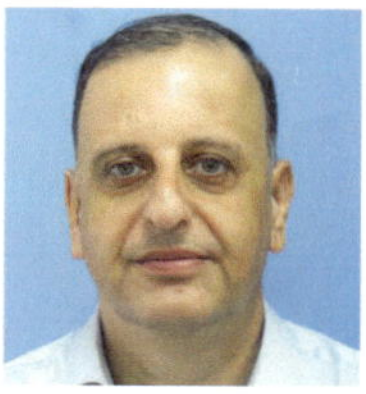

Dr. Uday M. Basheer Al-Naib is a senior lecturer at Universiti Teknologi Malaysia. He completed a postdoctorate in Materials Engineering at Universiti Sains Malaysia (USM) between 2013 and 2014. He received his Ph.D. in Materials Engineering from USM in 2013 and obtained his master's and bachelor's degrees in chemical engineering from the University of Baghdad and the University of Technology, Iraq, respectively. He has more than 30 years of academic and practical experience in the field of ceramic and metal composites, metal alloys, advanced ceramics, and ceramic–metal joining. He is the editor of a book series from IntechOpen. In addition, he is a recognized reviewer for Elsevier. He is acknowledged at the practical level for solving specific problems related to materials industries. This was evident when he published his research work with different publishers and in international material engineering journals with high impact factors. In addition to his own research, Dr. Uday has been acting as a supervisor of several academic theses in different fields of materials engineering. He received his CEng from the Engineering Council, UK, and became a professional member of the Welding Institute in May 2019.

Contents

Preface

In recent decades, the field of materials science has witnessed a surge in the exploration of advanced ceramics, leading to the development of novel materials with remarkable properties and diverse applications. Among these, zirconia has emerged as a prominent player, captivating the attention of researchers, engineers, and professionals across various industries. The allure of zirconia lies not only in its inherent structural and mechanical properties but also in its exceptional versatility that extends its application horizon from health care to electronics. This book, *Zirconia – New Advances, Structure, Fabrication and Applications*, is a comprehensive compilation of cutting-edge knowledge about zirconia, its advancements, structural intricacies, fabrication techniques, and its wide-ranging applications. Each chapter is thoughtfully crafted to provide a deeper understanding of zirconia, making this book an indispensable resource for students, researchers, practitioners, and anyone intrigued by the world of advanced ceramics.

The journey begins with an introductory chapter (Chapter 1) that lays the foundation for comprehending the significance of zirconia in modern materials science. This chapter offers a panoramic view of zirconia's historical evolution, its unique properties, and an overview of its myriad applications across different sectors. In Chapter 2, the focus shifts to a detailed exploration of the classification and generations of zirconia. The evolution of zirconia-based materials is traced through different phases and compositions, highlighting the progress made in enhancing its mechanical and optical properties. Chapter 3 delves into the fundamental processes of synthesizing zirconia, elucidating various methods such as solution combustion synthesis, sol-gel synthesis, co-precipitation, and hydrothermal and solid-phase sintering techniques. The chapter also extensively covers the characterization tools and techniques employed to analyze the structural, morphological, and functional properties of zirconia materials.

Chapter 4 embarks on a journey through the diverse processing techniques employed in shaping zirconia materials. From powder compaction to sintering and beyond, this chapter offers insights into the intricate procedures that transform raw zirconia into functional components with tailored properties. Zirconia's emergence as a dental biomaterial has revolutionized the field of restorative dentistry. Chapter 5 delves into the unique attributes of zirconia that make it an ideal candidate for dental applications, including its biocompatibility, mechanical properties, and aesthetic appeal. Building upon the foundation laid in the previous chapter, Chapter 6 takes a closer look at the specific dental applications where zirconia has found substantial traction. From dental crowns to bridges and implants, the versatility of zirconia in enhancing oral health care is thoroughly explored. The book culminates with an in-depth analysis of yttria-stabilized zirconia (YSZ), a prominent subcategory of zirconia. Chapter 7 elucidates the remarkable properties of YSZ that have led to its adoption in various high-tech applications, including solid oxide fuel cells, thermal barrier coatings, and more.

We extend our sincere gratitude to IntechOpen for their enthusiastic and professional support in bringing this book to fruition. Their dedication to disseminating knowledge has played a pivotal role in making this compilation a reality. As you embark on this journey through the multifaceted world of zirconia, we hope that the insights and revelations offered within these pages inspire further innovation and exploration in the realm of advanced ceramics.

Dr. Uday M. Basheer Al-Naib
Centre for Advanced Composite Materials (CACM),
Faculty of Mechanical Engineering,
Universiti Teknologi Malaysia,
Johor Bahru, Malaysia

Section 1

Introduction to Zirconia

Chapter 1

Introductory Chapter: Introduction to Zirconia Ceramic – A Versatile and Durable Material with a Wide Range of Applications

Uday M. Basheer Al-Naib

1. Introduction

Zirconium oxide, also known as zirconia, is a ceramic material that has many important properties, including high strength, good toughness, and excellent corrosion resistance [1, 2]. It is a white crystalline material made from the mineral zircon found in various parts of the world. Zirconia exhibits numerous attributes that hold significance within industrial applications including use in the manufacture of ceramics, abrasives, and refractories, and as a structural material in the aerospace and automotive industries [3]. In addition to its practical uses, zirconia is also known for its beautiful, diamond-like appearance, which has made it a popular use in jewelry and other decorative items [4].

Zirconia possesses a multitude of distinct properties that render it a compelling material for a diverse range of applications [5]. With an exceptionally elevated melting point, zirconia demonstrates remarkable resistance to elevated temperatures and thermal shocks. Furthermore, its exceptional performance as an electrical insulator and its minimal coefficient of thermal expansion contribute to its resilience against thermal stresses. Beyond these qualities, zirconia's notable strength and hardness make it a prime candidate for structural applications. Moreover, its pronounced corrosion resistance positions it as a superb option for deployment in demanding environmental conditions [3].

When it comes to its manufacturing process, zirconia is created by subjecting zircon sand to exceptionally elevated temperatures within a specialized furnace. This procedure yields zirconium oxide as the final product [6]. The technique employed is known as sintering, which yields a compact and durable substance amenable to various configurations. Once crafted, zirconium oxide finds diverse applications based on particular traits and needs. Notably, it finds utility in crafting cutting implements and components resistant to wear, and it serves pivotal roles within the aerospace, automotive, and medical sectors. Its biocompatibility renders it valuable for producing medical equipment, including dental implants and other healthcare devices [7].

2. History of zirconia

Zirconia possesses a captivating and extensive historical background. Its origins trace back to the eighteenth century, marked by the pioneering work of German scientist Martin Heinrich Klaproth. He succeeded in extracting zirconia from the mineral zircon, an enduringly precious gemstone with a centuries-old legacy [6, 8]. In the nineteenth century, zirconia was first used as an abrasive, and it was later found to possess several unique properties that made it useful for a wide variety of applications. One of the earliest uses of zirconia was in the manufacture of abrasives and it was used to make grinding wheels and other abrasive tools. In the twentieth century, zirconia began to be used in ceramic manufacture, and proved to be an excellent material for use in structural applications due to its high strength and toughness. Zirconia has also been shown to have excellent corrosion resistance and a low thermal expansion coefficient, making it an ideal material for use in harsh environments. Zirconia has continued to be a popular and widely used material in the twenty-first century [9]. In the medical field, zirconia has been used to manufacture dental implants, as well as other medical devices such as prosthetics and surgical tools. In the aerospace industry, zirconia has been used as a structural material due to its high strength and corrosion resistance. It has also been used in the production of cutting tools, wear-resistant parts, and other industrial components [10, 11].

3. Structure of zirconia

Zirconia displays polymorphism across varying temperatures, manifesting itself in three distinct shapes (**Figure 1**): monoclinic or baddeleyite (from room temperature up to 1170°C), tetragonal (1170–2370°C), and cubic (2370–2700°C, aligning with its melting point) [12]. At room temperature, zirconia exists in a monoclinic crystalline structure, but it can also exist in other crystalline structures depending on the temperature and pressure conditions. One of the unique properties of zirconia is that it can undergo a phase transition from monoclinic structure to a tetragonal structure under severe stress [13]. This alteration in phase brings about heightened toughness and ductility in the material, making it an ideal material for use in structural applications where it may be subjected to high stresses. Beyond its crystalline arrangement, zirconia's notable attributes encompass elevated density and hardness, qualities

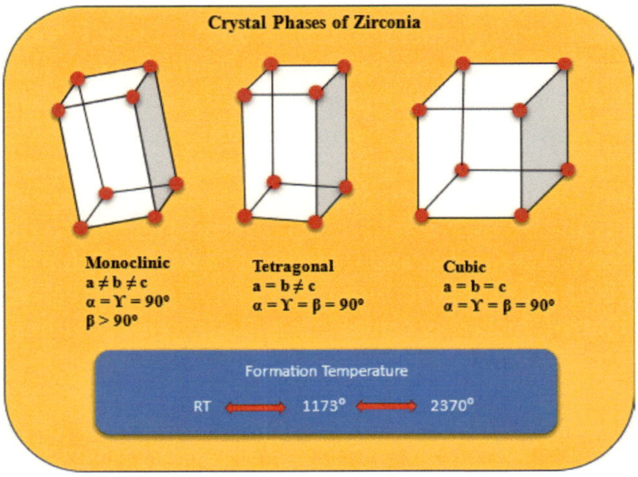

Figure 1.
Structure of zirconia.

attributed to robust chemical bonds. These characteristics collectively bestow zirconia with remarkable potency and resilience, positioning it as an optimal choice across an extensive array of applications [14].

Zirconia boasts an intricate crystalline arrangement, composed of zirconium oxide molecules arranged in a repetitive pattern [15]. This material exhibits refractory qualities, indicating a remarkable resistance to chemical reactions and an elevated melting point. At ambient conditions, zirconia takes on a monoclinic crystalline structure, characterized by the intersection of two axes at an angle. This configuration remains stable at room temperature and is marked by low density and strength. Yet, under substantial loads, zirconia can undergo a reversible phase transition, shifting from monoclinic to tetragonal. This phase transition is reversible and occurs when the material is stressed beyond a certain threshold [16]. The tetragonal structure exhibits superior density and strength compared to the monoclinic form, and it also demonstrates greater ductility, rendering it more resilient against cracks and fractures when burdened. Beyond its crystalline arrangement, zirconia is distinguished by high density and hardness, attributes attributed to robust chemical bonds within the material. These intrinsic characteristics bestow upon zirconia exceptional strength and endurance, rendering it an optimal choice for a wide array of applications. Moreover, zirconia functions as an outstanding electrical insulator and exhibits a low coefficient of thermal expansion, which safeguards it against thermal strains and positions it as a prime candidate for high-temperature environments [17]. Collectively, zirconia's intricate structure bequeaths it with a distinctive amalgamation of traits, establishing its significance across diverse industries.

Zirconia's distinct attributes are also influenced by a range of structural properties [18]. These encompass its microstructure, grain size, and defects. A material's microstructure consists of the arrangement of its atoms and their spatial relationships. In the case of zirconia, its microstructure can exhibit variability contingent upon the specific production conditions employed. For instance, when zirconia is manufactured via sintering, its microstructure often assumes a more porous nature, yielding a heightened surface area. Conversely, the application of hot isostatic pressing yields zirconia with a denser microstructure and a diminished surface area [19]. Notably, zirconia's microstructure holds considerable sway over its attributes, including strength and toughness [20].

4. Types of zirconia

There are different types of zirconia used in different applications. The most common types of zirconia include: (1) Monoclinic zirconia: This is the most stable form of zirconia at room temperature and is widely used in grinding and cutting tools, as well as in ceramic and refractory applications, (2) Tetrahedral zirconium: This form of zirconium undergoes a phase transition from monoclinic to tetragonal at a certain temperature known as the "transition temperature" [21]. Quadrilateral zirconia has higher strength and toughness than monoclinic zirconia, making it suitable for applications where these properties are important, (3) Cubic zirconia: This type of zirconia has a high refractive index and is often used as a substitute for diamond in jewelry. It is also used in high temperature thermal protection tubes and other grinding and high temperature applications, and (4) Stable zirconia: This type of zirconia is made by adding small amounts of other elements such as yttrium or magnesia to improve the stability and strength of the material. Stabilized zirconia is widely used in abrasive and high temperature applications, as well as in fuel cell electrodes and other advanced technologies. The type of zirconia used in a particular application depends on the specific properties and performance requirements for that application [22].

5. Production of zirconia

Zirconia, also referred to as zirconium dioxide, is synthesized through a procedure known as calcination. This process entails subjecting natural zircon, a mineral containing zirconium, to elevated temperatures in the presence of oxygen [23]. This chemical reaction transforms zircon into zirconia (**Figure 2**). The initial stage in zirconia production involves the extraction of zircon from natural deposits, usually undertaken within mining operations that involve excavating the earth to access zircon reservoirs [25]. Once extracted, zircon is purified to eliminate impurities. The refined zircon is then finely ground and combined with other substances like alumina or magnesia to generate a uniform mixture. This blend is introduced into a calcination furnace, where it is subjected to high temperatures (typically around 2000°C) in an oxygen-rich environment. During this calcination process, zircon reacts chemically with oxygen, resulting in its transformation into zirconia. Subsequently, the zirconia powder formed is cooled and collected for subsequent procedures [24]. Following

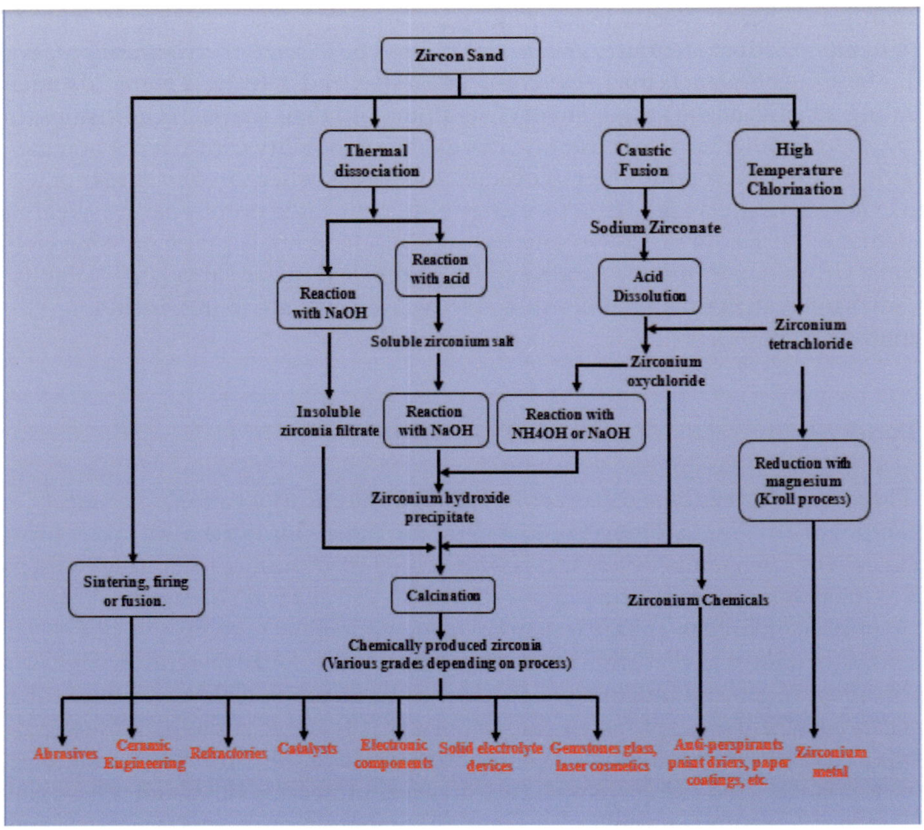

Figure 2.
The production processes of zirconia [24].

calcination, additional processing steps, such as pressing and sintering, may be applied to the zirconia powder. These steps help shape the material into its desired form while achieving specific properties. Eventual zirconia products find application across a broad spectrum of uses, encompassing abrasives, ceramics, refractories, and cutting-edge technologies.

Multiple techniques exist for producing zirconia, with their suitability contingent upon the distinctive attributes and performance prerequisites of the product. Further insights into zirconia production encompass:

Following calcination, the zirconia powder can undergo additional refinement via "hot pressing" or "hot isostatic pressing." This technique entails pressing the zirconia powder under elevated pressure and temperature, resulting in a compact, robust structure with heightened strength and toughness. An alternate method, termed "sintering," involves subjecting the zirconia powder to elevated temperatures (typically around 1600 to 1800°C) in a dedicated sintering furnace. As the powder heats, its particles coalesce, yielding a consolidated piece characterized by improved density and strength. Beyond hot pressing and sintering, another avenue for zirconia processing is "reaction bonding." In this approach, zirconia powder is blended with a metal oxide like alumina or magnesia, and the mixture is then subjected to high temperatures. The ensuing chemical reaction establishes a bond between the zirconia and the metal oxide, yielding a sturdy, fortified piece with heightened strength and toughness [26]. The selection of a specific zirconia production method hinges upon the distinct attributes and performance requisites of the final product [27]. Considerations encompass parameters like density, strength, toughness, and various other properties that influence the optimal manufacturing route.

6. Future developments in zirconia technology

There are several areas of research and development that are focused on advancing the technology of zirconia. Some potential future developments in zirconia technology include [28]:

1. Enhanced Strength and Toughness: Scientists are exploring novel processing techniques and additives to heighten the strength and toughness of zirconia. This endeavor aims to render it more apt for applications necessitating structural integrity and load-bearing capacities.

2. Augmented Corrosion Resistance: While zirconia boasts exceptional corrosion resistance, researchers are investigating avenues to further enhance this trait. This could potentially involve the development of fresh coatings or innovative additives.

3. Elevated Electrical and Thermal Conductivity: Zirconia's prowess as an electrical insulator is recognized, yet researchers are delving into additives or doping methods to elevate its electrical conductivity for specific applications. Parallelly, efforts are underway to bolster zirconia's thermal conductivity, perhaps through the incorporation of conductive additives or novel processing techniques.

4. Biomedical Innovations: Zirconia holds promise in the realm of biomedicine, potentially finding utility in medical implants and prosthetics. Researchers are directing their efforts toward optimizing zirconia's surface attributes and biocompatibility for these medical applications.

7. Conclusion

Zirconia ceramic emerges as a resilient and versatile substance, renowned for its multifaceted properties such as elevated melting point, robust strength, notable toughness, minimal thermal expansion, chemical resilience, and remarkable wear resistance. This amalgamation of attributes equips it for a diverse array of applications, spanning industrial and biomedical domains. Its utility encompasses cutting tools, wear-resistant coatings, dental prosthetics like crowns and bridges, refractory components, catalysts, and abrasives. Furthermore, the burgeoning interest in leveraging zirconia within fuel cell technology is fuelled by its aptitude for oxygen ion conduction at elevated temperatures. Overall, zirconia stands as a highly advantageous and invaluable material, exerting its influence across an extensive spectrum of fields.

Author details

Uday M. Basheer Al-Naib
Centre for Advanced Composite Materials (CACM), Faculty of Mechanical Engineering, Universiti Teknologi Malaysia, Malaysia

*Address all correspondence to: ummb2008@gmail.com

References

[1] Kalavathi V, Bhuyan RK. A detailed study on zirconium and its applications in manufacturing process with combinations of other metals, oxides and alloys—A review. Materials Today: Proceedings. 2019;**19**:781-786

[2] Qi B et al. ZrO2 matrix toughened ceramic material-strength and toughness. Advanced Engineering Materials. 2022;**24**(6):2101278

[3] Vasile BS et al. Ceramic composite materials obtained by electron-beam physical vapor deposition used as thermal barriers in the aerospace industry. Nanomaterials. 2020;**10**(2):370

[4] Hurrell K, Johnson ML. Gemstones: A Complete Color Reference for Precious and Semiprecious Stones of the World. New York, USA: Chartwell Books; 2016

[5] Correia DM et al. Ionic liquid–polymer composites: A new platform for multifunctional applications. Advanced Functional Materials. 2020;**30**(24):1909736

[6] Arnold B. Zircon, Zirconium, Zirconia-Similar Names, Different Materials. Berlin, Germany: Springer; 2022

[7] Fathi R et al. Past and present of functionally graded coatings: Advancements and future challenges. Applied Materials Today. 2022;**26**:101373

[8] Halka M, Nordstrom B. Transition Metals. New York, USA: Infobase Holdings, Inc; 2019

[9] Mogensen MB. Materials for reversible solid oxide cells. Current Opinion in Electrochemistry. 2020;**21**:265-273

[10] Okada A. Automotive and industrial applications of structural ceramics in Japan. Journal of the European Ceramic Society. 2008;**28**(5):1097-1104

[11] Hoornaert T, Hua Z, Zhang J. Hard wear-resistant coatings: A review. In: Advanced Tribology: Proceedings of CIST2008 & ITS-IFToMM2008. China; 2010. pp. 774-779

[12] Bocanegra-Bernal M, De La Torre SD. Phase transitions in zirconium dioxide and related materials for high performance engineering ceramics. Journal of Materials Science. 2002;**37**:4947-4971

[13] Zhang N, Zaeem MA. Competing mechanisms between dislocation and phase transformation in plastic deformation of single crystalline yttria-stabilized tetragonal zirconia nanopillars. Acta Materialia. 2016;**120**:337-347

[14] Geetha M et al. Ti based biomaterials, the ultimate choice for orthopaedic implants–A review. Progress in Materials Science. 2009;**54**(3):397-425

[15] Rushton MJ et al. Stoichiometry deviation in amorphous zirconium dioxide. RSC Advances. 2019;**9**(29):16320-16327

[16] Swain M. Shape memory behaviour in partially stabilized zirconia ceramics. Nature. 1986;**322**(6076):234-236

[17] Xu L et al. A new class of high-entropy fluorite oxides with tunable expansion coefficients, low thermal conductivity and exceptional sintering resistance. Journal of the European Ceramic Society. 2021;**41**(13):6670-6676

[18] El-Wazery M, El-Desouky A. A review on functionally graded ceramic-metal materials. Journal of Materials and Environmental Science. 2015;**6**(5):1369-1376

[19] Mayo MJ. Processing of nanocrystalline ceramics from ultrafine particles. International Materials Reviews. 1996;**41**(3):85-115

[20] Rathee G et al. Emerging multimodel zirconia Nanosystems for high-performance biomedical applications. Advanced NanoBiomed Research. 2021;**1**(9):2100039

[21] Gorelov V. High-temperature phase transitions in ZrO 2. Physics of the Solid State. 2019;**61**:1288-1293

[22] Zhao W et al. High-temperature oxidation behavior of Zr-4 and Zr-Sn-Nb alloy in different oxidation ambient. Journal of Alloys and Compounds. 2021;**887**:161396

[23] Ende M et al. Dry annealing of radiation-damaged zircon: Single-crystal X-ray and Raman spectroscopy study. Lithos. 2021;**406**:106523

[24] Trevisi R et al. Radiological protection in industries involving NORM: A (graded) methodological approach to characterize the exposure situations. Atmosphere. 2023;**14**(4):635

[25] Liu H. Investigation on joint mining of rare earth elements and zircons from the Yangtze River, China. In: Journal of Physics: Conference Series. Hangzhou, China: IOP Publishing; 2021

[26] Bapat RA et al. Review on synthesis, properties and multifarious therapeutic applications of nanostructured zirconia in dentistry. RSC Advances. 2022;**12**(20):12773-12793

[27] Katoh Y et al. Properties of zirconium carbide for nuclear fuel applications. Journal of Nuclear Materials. 2013;**441**(1-3):718-742

[28] Olhero S et al. Conventional versus additive manufacturing in the structural performance of dense alumina-zirconia ceramics: 20 years of research, challenges and future perspectives. Journal of Manufacturing Processes. 2022;**77**:838-879

Section 2

Classification of Zirconia

Chapter 2

Classification and Generations of Dental Zirconia

Ali Dahee Malallah and Nadia Hameed Hasan

Abstract

Zirconium oxide (ZrO2) is polymorphic (temperature dependent) structure; zirconia can take three crystallographic forms at ambient pressure. Under normal conditions, pure zirconia is monoclinic (m). At (1170°C), the substance converts to a tetragonal crystal structure (t), then to a cubic crystal structure (c) at (2370°C), and finally to a fluorite structure above (2370°C), melting at (2716°C). During the heating and cooling cycles, the Zirconium oxide ceramic undergoes a hysteretic, martensitic t- m transformation, which is reversible at 950°C upon cooling. For dental applications, various types of zirconium-dioxide (zirconia) materials are available. These materials have a variety of chemical compositions, crystal configurations, manufacturing processes, and important variations in their mechanical and optical properties. Numerous generations of zirconia materials have been developed, ranging from the use of zirconia crystals as reinforcement elements in zirconia toughened alumina (ZTA) to partially stabilized zirconia (PSZ) and the conventional (3Y-TZP) to the appearance of new translucent zirconia materials such as cubic stabilized zirconia (CSZ).

Keywords: zirconia, monolithic, tetragonal polycrystal zirconia, partially stabilized zirconia, zirconia in dentistry

1. Introduction

Zirconium (Zr) is a chemical element whose name Zr is taken from the name of mineral zircon and the name zircon comes from the Persian word "Zar-Gun," which means "golden color".Zr is a transition metal with an atomic mass of (91.224 g/mol) and an atomic number of (40). The melting point of Zr is (1855°C), while the boiling point is (4371°C). Zr was discovered in 1789 by Martin Heinrich Klaproth, a German scientist and isolated in 1824 by Swedish chemist Jöns Jacob Berzelius (Gautam *et al.*, 2016; Nistico, 2021). Helmer and Driskell reported the first biomedical application of Zr in 1969; however Christel (1988) was the first to use Zr to make a ball head for a complete hip replacement [1, 2].

Zr is never found in nature as a naturally occurring metal. It occurs naturally in igneous rocks in combination with other elements such as iron, titanium, and silicon oxide. The most abundant source of Zr is zircon (ZrSiO4), which is found mostly in Australia, South Africa, Brazil, India, Russia, and the United States. Many other mineral species contain Zr, including baddeleyite. Despite the fact that zirconia was first used in orthopedics in 1969 for hip head replacement, it was not used in dentistry until the 1990s. Zirconia-based ceramics such as tetragonal zirconia which is partially stabilized by yttria have been successfully incorporated into daily dental work to construct fixed dental prostheses (FDPs) and dental (CAD/CAM)system based on their excellent biological, mechanical, and physical properties. Many endeavors, on the other hand, have undergone numerous improvements in composition and microstructure in order to improve their optical properties without minimizing their mechanical properties [3–5].

Zirconia is a non-etchable polycrystalline material as it has no glassy phase within its structure that can be bonded to tooth structures using both traditional and resin cements; however, resin cements are favored because they have a great marginal seal, better retention, and improved ceramic fracture resistance. Zirconia –based ceramics are conventionally used as substructure materials that require a veneer facing for clinically acceptable appearance due to their high opacity and whitish optical appearance with excellent biocompatibility and lower the risk of pulp irritation that may occur due to the lower thermal conductivity. Ceramic facing, on the other hand, has some drawbacks, such as poor tensile strength and crack toughness. Porcelain is vulnerable to cracking when stressed due to intrinsic flaws in the crystal structure and the presence of voids. As a result, veneer chipping has been described as a major cause of failure and the most common complication in all-ceramic crowns [6, 7].

A new generation of zirconia, known as "monolithic zirconia," has recently been introduced. Monolithic zirconia restorations, according to the manufacturers, has a higher translucency than conventional zirconia and therefore does not need a veneer layer. Monolithic zirconia restorations can have a number of clinical advantages. Without the use of a veneer, the amount of tooth loss and the risk of chipping are reduced, as well as the restorations' actual strength will be increased [8, 9].

2. Classification of zirconia

Over the past 20 years, there have been more varieties of dental zirconia, making it occasionally challenging to select the right kind for each restoration. Utilizing CAD/CAM technologies, it is now possible to create dental restoratives with extremely precise fitting owing to advancements in digital technology. Metal-free restorations are also appealing for biological and cosmetic reasons. For dental applications, various types of zirconium-dioxide (zirconia) materials are available. These materials have a variety of chemical compositions, crystal configurations, manufacturing processes, and important variations in their mechanical and optical properties. Numerous generations of zirconia materials have been developed, ranging from the use of zirconia crystals as reinforcement elements in zirconia toughened alumina (ZTA) to partially stabilized zirconia (PSZ) and the conventional (3Y-TZP) to the appearance of new translucent zirconia materials such as cubic stabilized zirconia [10, 11]. **Figure 1** is showing recent classification of yttria stabilized zirconia.

Zirconia can be classified according to the following criteria into:

2.1 Zirconia microstructure

The monoclinic phase of pure zirconia (ZrO2) is stable at ambient temperature, while the tetragonal and cubic crystal phase systems change with temperature. When

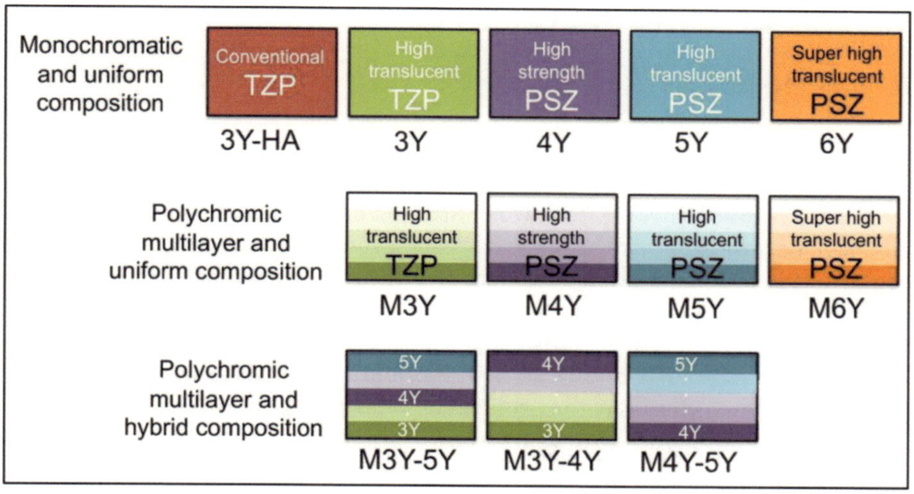

Figure 1.
Structural schematic diagram and classification of yttria-stabilized dental zirconia [10].

zirconia is solidly dissolved in yttrium(Y), calcium (Ca), magnesium (Mg), cerium (Ce), or other ions with an ionic radius larger than that of zirconium (Zr), and accordingly divided into [12]:

2.1.1 Partially stabilized zirconia (PSZ)

The tetragonal and cubic phase systems become stable at room temperature when zirconia is solidly dissolved in yttrium (Y), calcium (Ca), magnesium (Mg), cerium (Ce), or other ions with a greater ionic radius than that of zirconium (Zr) [1, 13]. Cubic-stabilized zirconia (CSZ) is the cubic phase zirconia that is stable at room temperature when yttria (Y2O3) is introduced in amounts greater than 8 mol%. Tetragonal and cubic phases are mixed at room temperature when yttria is 3 to 8 mol%, and this material is known as partially stabilized zirconia (PSZ). Tetragonal zirconia polycrystal (TZP), also known as toughened zirconia, is zirconia that is close to 100% tetragonal at room temperature when yttria is approximately 3 mol%. This 3 mol% yttria tetragonal zirconia polycrystal (3Y-TZP) was one of the earliest zirconia used as “white metal” in dentistry [10].

2.1.2 Fully stabilized zirconia (FSZ)

It is a cubic zirconia that includes no less than 8% yttrium oxide(Y2O3), hence the term “fully” stabilized zirconia (FSZ). presenting tetragonal and cubic crystals on its microstructure, being totally inert to the aging in autoclave because cubic zirconia does not exhibit the t/m transformation toughening phenomenon. The volume of cubic crystals is greater. They are less firmly bonded and, as a result, have increased light scattering at the grain boundaries due to reduced residual porosity. Additionally, incident light is emitted more uniformly in all spatial directions because the cubic crystal structures are more isotropic than the tetragonal structures. To increase the translucency of dental zirconia, several producers have recently started producing it in its full cubic stabilized form (due to increased cubic phase) [14–17].

2.2 Zirconia used with porcelain facing or not

It can be divided into:

2.2.1 Monolithic (full contour restoration)

The terms "monolithic" is Greek word that means: mono is "single" and "lithos" is "stone," This indicates that the materials have a consistent appearance. As microstructures, monolithic materials consist of two or more phases such as in zirconia it has three phases monoclinic, tetragonal and cubic phases. With the continuous development of new and more translucent Y-TZP and the advancement of CAD-CAM technology that facilitate the fabrication of monolithic Y-TZP crowns and FDPs. These systems aim to eliminate the problem of veneering porcelain chipping and provide acceptable esthetics with characterization, which may be an esthetic option in the molar area without a reduction in strength. Full-contour zirconia restorations, however, do not have adequate translucency because the matrix and zirconia particles have different refractive indices. To produce restorations with great translucency and to achieve zirconia's mechanical qualities, a number of brands of monolithic zirconia have been introduced. To increase zirconia's translucency, a full cubic stabilized monolithic zirconia (FSZ) has been produced. Comparing newly developed Y-TZP monolithic materials to traditional zirconia, the translucency has enhanced and low temperature degradation has been limited due to the materials' lower alumina content, relatively fine grain size, and presence of optically isotropic cubic zirconia particles. A monolithic multilayer zirconia has been developed. It is a polychromatic, translucent zirconia with combined shade and translucency gradient [18–20].

2.2.2 Core build up (veneered with porcelain facing)

A high strength ceramic substructure made of zirconia or alumina is the foundation of bi-layered crowns, which are then veneered with ceramic or dental porcelain such as feldspathic porcelain. Despite having good esthetic qualities, the resulting restorations are prone to failure, such as chipping of the veneering ceramic. An alternative approach is to eliminate the veneer and produce a full contour monolithic zirconia crown but the possibility of wearing the opposing natural teeth is still a concern. Furthermore, the potential loss of strength brought on by low temperature aging or degradation (LTD), which may be induced in an aqueous environment. By completely covering the zirconia restoration with a ceramic veneer, it can prevent it from coming into direct contact with the oral cavity and potentially prevent this occurrence [20, 21].

2.3 Colored and non-colored zirconia

Dental zirconia has changed over the past 20 years from its original white, opaque appearance to translucent and chromatic as well as polychromatic (multi-layered) forms. Zirconia has become as the most versatile restorative material, offering a wide spectrum of translucency and colors.

2.3.1 Non dyed (monochromatic white)

Non dyed blanks have a hard-white monochrome color, which can be an esthetic disadvantage in many indications. The restorations milled in the white body condition can be manually and individually colored with coloring oxides after the milling process and sintered afterwards to overcome this drawback. The form-milled open-pore framework is immersed in the suitable colored liquid for a brief period of time to dye it. Alternatively, brushes can be used to create color gradients that are comparable to various colored liquids of varying intensities. The sintering procedure is performed after removing the excess remaining color while it is still wet and drying the framework. Utilizing liner or stain is another way to make white zirconia more esthetically beautiful [17].

2.3.2 Dyed polychromatic (multilayer)

Construction of polychromatic zirconia in the form of a gradational multilayered zirconia disc with two to seven or even more layers of color, mimicking the appearance of layered porcelain in full-contour monolithic restorations, and the creation of zirconia blanks with multiple, different shaded layers make full-contour zirconia restorations more esthetically pleasing than those made with conventional monochromatic zirconia. After sintering, full-contour zirconia restorations with esthetics that outperform monochromatic restorations can be obtained. The Katana Multi-Layered zirconia by Kuraray Noritake Dental Japan was the first polychromatic zirconia to be introduced to the dental market in February 2015. They are intended for the production of full contour zirconia crowns with a greater level of esthetics. Since then, virtually every significant dental company has introduced their own version of multi-layer zirconia [20, 22].

2.4 Fully sintered or partially sintered zirconia blanks

2.4.1 Fully sintered or hot isostatic procedure (HIP)

Performed by a hot isostatic press at 1000 bar and 50°C below the sintering temperature at 1400–1500°C under high pressure and an inert gas atmosphere to reduce the material porosity and ensure high values of toughness and translucency of zirconium ceramics by such procedure. Carrying out HIP on Y-TZP results in a gray-black material that usually requires subsequent heat treatment to oxidize and restore whiteness. Then, a specially designed milling system used to machine the blanks, but with a low machinability and high hardness of a fully sintered Y-TZP, such milling system has to be particularly robust [23, 24].

2.4.2 Partially sintered or cold isostatic pressing

With cold isostatic pressing, the powders are shaped into ceramic blanks. The powder of the partially sintered blanks typically contains a binder that enhance its pressing, but during the step of a pre-sintering it will be eliminated. The most widely used procedural method for Y-TZP shaping is cold isostatic pressing, which yields stable, chalk-like non-sintered green-stage objects with a very high primary density. Sintering without pressure in the oxidizing atmosphere of a specific furnace allows the green objects to be further stabilized and condensed up to approximately 95% of their theoretical density., the pre-sintering temperature of the heat treatment effects on the machined blank roughness, a higher pre-sintering temperature gives a rougher surface, so the choice of a proper pre-sintering temperature will be thus critical. These materials remain softer than the HIP zirconia and are easier to mill [24].

2.5 Zirconia generations

2.5.1 First generation

The first generation of 3 mol% yttria (3Y-TZPs) had flexure strengths greater than 1 GPa and contained 0.25 wt% alumina (Al2O3) as a sintering aid. However, that zirconia displayed considerable opacity, due to the inherent birefringence of noncubic zirconia phases as light is scattered at the grain boundaries, pores, and additional inclusions. They were primarily used as framework materials in posterior and anterior fixed dental prostheses (FDPs) and porcelain-veneered crowns [25, 26].

2.5.2 The second generation

The alumina content of the second iteration of 3 mol% yttria (3Y-TZP), used in dentistry was reduced from 0.25 wt% to 0.05 wt%. The 3Y-TZP with 0.05 wt% alumina is more translucent than the 3Y-TZP with 0.25 wt% alumina, but because there is less alumina to stabilize the tetragonal phase, it is more prone to low-temperature degradation. Some confusion in nomenclature has occurred as both 0.05 wt% alumina- containing 3Y-TZP and 5 mol% yttria-stabilized zirconia polycrystal (5Y-ZP) have been called "translucent zirconia"; however, these zirconia materials have different mechanical and optical properties [26, 27].

2.5.3 The third generation

The third iteration of dental zirconia is doped with 5 mol% yttria, which creates a partially stabilized zirconia with around 50% cubic phase zirconia is produced. Zirconia's cubic phase is isotropic in many crystallographic directions, which reduces the amount of light scattering at grain boundaries. The cubic zirconia thus seems more transparent. Since stabilized cubic zirconia does not change at room temperature, it will not degrade at low temperatures or undergo transformation toughening. It has diminished mechanical qualities but will not change with time [26].

2.6 Zirconia according to mol% concentration of yttria

2.6.1 3 mol% yttria-partially stabilized zirconia (3Y-PSZ)

Mechanical and optical properties of zirconia depend on the yttria mol% content. Of all the several forms of zirconia, 3Y-PSZ has the greatest values for opacity, flexural strength, and fracture toughness. The material's high color value and opacity restrict its use to posterior restorations [26].

2.6.2 4 mol% yttria-partially stabilized zirconia (4Y-PSZ)

4Y-PSZ zirconia has translucency and mechanical properties in between 3Y-PSZ zirconia and 5Y- PSZ zirconia making it an attractive in between material for esthetic zone [26].

2.6.3 5 mol% yttria-partially stabilized zirconia (5Y-PSZ)

5Y-PSZ zirconia has enhanced translucency, but reduced mechanical properties [26]. **Table 1** is showing zirconia by the yttrium content and its effect on physical and mechanical properties.

3. Zirconia ceramics for dental applications

The following are the three most common forms of zirconia used in dentistry today [1]:

a. Tetragonal zirconia polycrystals doped with yttrium cation (3Y TZP)

Traditional 3Y-TZP zirconia is the most commonly used zirconia material in bilayered dental restorations as a substitute for the metal substructure. The 3Y-TZP is the toughest zirconia material currently available in dentistry. As a stabilizer, 3Y-TZP contains 3 mol% Yttria (Y_2O_3). The stabilizer Y^{3+} cations and Zr^{4+} ions are distributed randomly over the cationic sites, but electrical neutrality is attained by creating oxygen vacancies [29].

The mechanical properties of 3Y-TZP are highly influenced by the particle size. 3Y-TZP becomes less vulnerable to spontaneous tetragonal to monoclinic (t-m) transformation when the particle size is less than (1 μ). Furthermore, the transformation is not possible below a certain particle size (0.2 μ), resulting in decreased fracture toughness. A particle size greater than (1 μ), on the other hand, makes 3Y-TZP less stable and more vulnerable to spontaneous t-m transformation, resulting in huge volume increase and a decrease in fracture toughness. As a result, it appears that the critical particle size for 3Y-TZP is 1(μ) [30].

According to this, the sintering conditions that result in an increased particle size have a significant effect on the final ceramic product's stability and strength. Larger particle sizes result from higher sintering temperatures and longer sintering times. Depending on the producer, the final sintering temperatures of 3Y-TZP range from (1350 to 1550C). Furthermore, the particle size and phase stability of 3Y-TZP used in dental applications can be affected by variations in sintering temperatures during firing [30–32].

b. Partially stabilized zirconia doped with magnesium cation (Mg-PSZ)

The quantity of magnesium oxide in commercial materials ranges from 8 to 10 mol%. To control the fracture toughness of Mg-PSZ, high sintering temperatures (1680 and 1800 C) and tightly controlled cooling cycles are required. Because of the less stable molecular composition of this material, early t-m transition may occur, leaving insufficient tetragonal zirconia for the material to transform and toughen upon further fracture formation. Lower mechanical properties are the outcome of the higher prevalence of the monoclinic form. Due to the existence of surface porosity associated with a large particle size (30–60 μ), magnesium stabilized zirconia (Mg-PSZ) has not been widely used in biomedical applications due to the risk of excessive wear [1].

3Y-TZP: Opaque Zirconia	4Y-TZP: Some Translucency	5Y-TZP: Most Translucency
High mechanical properties	High mechanical properties	High mechanical properties
White opaque	Some Translucency	4Y-TZP: High translucency
Low temperature degradation	Decreased low temperature degradation	Little or no low temperature degradation
Mainly tetragonal phase	Tetragonal and cubic phases	More Cubic less tetragonal phases

Table 1.
Defining zirconia used in dentistry by the yttrium content and its effect on physical and mechanical properties [28].

c. Zirconia toughened alumina (ZTA)

Another way to take advantage of stress-induced zirconia transformation is to use zirconia toughened alumina (ZTA), which is composed of an alumina matrix and zirconia. The commercially accessible dental product In-Ceram Zirconia (VidentTM, Brea, CA) was developed by adding 33 vol.% of 12 mol % ceria-stabilized zirconia (12Ce-TZP) to In-Ceram Alumina. Slip casting or soft machining may be used to create In-Ceram Zirconia restorations. Increased porosity is likely related to In-Ceram zirconia's lower mechanical properties as compared to 3Y-TZP dental ceramics [33].

4. Zirconia microstructure

Zirconium oxide (ZrO_2) is polymorphic (temperature dependent) structure; zirconia can take three crystallographic forms at ambient pressure. Under normal conditions, pure zirconia is monoclinic (m). At (1170°C), the substance converts to a tetragonal crystal structure (t), then to a cubic crystal structure (c) at (2370°C), and finally to a fluorite structure above (2370°C), melting at (2716°C). During the heating and cooling cycles, the Zirconium oxide ceramic undergoes a hysteretic, martensitic t- m transformation, which is reversible at 950°C upon cooling [2, 34]. **Figure 2** is showing Crystallographic phases of zirconia and temperature hysteresis.

Monoclinic phase (m) is a deformed prism with parallelepiped sides that has lower mechanical properties and may lead to a reduction in the cohesion of ceramic particles [35].

Tetragonal phase (t): is a straight prism with rectangular sides that possesses increased mechanical properties. Precipitation of a finely dispersed tetragonal phase within a cubic matrix, capable of being changed into the monoclinic phase when the matrix's pressure was alleviated by a crack propagated within the matrix [35].

Cubic phase (c): in the shape of a square-sided straight prism has a large grain size, which reduces light scattering and birefringence at grain boundaries. The cubic phase of zirconia is isotropic in different crystallographic directions, which reduces light scattering at grain boundaries. The cubic zirconia becomes more transparent as a result. At room temperature, stabilized cubic zirconia does not transform, so it will not undergo transformation toughening or low-temperature

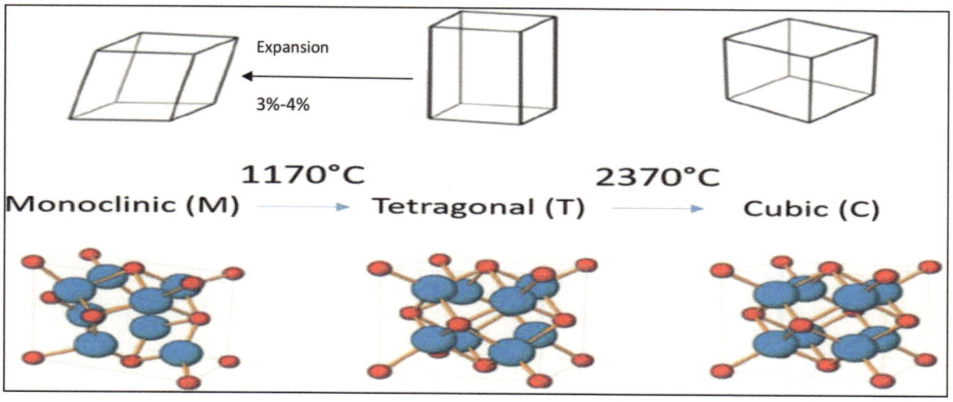

Figure 2.
Crystallographic phases of zirconia, temperature and hysteresis [2].

degradation. In other words, it has weakened mechanical properties but will not change over time [28, 35].

5. Generations of zirconia in dentistry

There are four main generation of zirconia in dentistry and new multilayer novel generation of zirconia:

i. **(First generation 3Y-TZP 0.25**Al_2O_3**):** this was introduced about two decades ago. Zirconia crystals are partly stabilized with 3 mol % yttria (4.5–6 percent by weight of Y_2O_3). Owing to the presence of alumina, it is opaque. When green-state zirconia is put in the furnace, alumina is applied as a sintering aid to help prevent the formation of pores [7, 35].

ii. **(Second generation 3Y-TZP 0.05** Al_2O_3): this was introduced between 2012 and 2013. The zirconia crystal structure is partly stabilized with 3 mol percent yttria (Y_2O_3), but altered by providing a lower volume of aluminum oxide (Al_2O_3) with smaller Al_2O_3 grain sizes and re-positioned in the zirconia matrix. Since there is a less alumina to stabilize the tetragonal phase, it is more prone to low-temperature degradation [28].

iii. **(Third generation 5Y-TZP 0.05** Al_2O_3): Between 2014 and 2015, a new category of translucent zirconia materials, also known as Anterior Zirconia, High Translucent Zirconia, or Cubic stabilized Zirconia (CSZ), was developed to overcome the translucency limitation. The translucency of the new materials is produced by increasing the yttria content from 9 to 10% by weight, which results in the production of 50% cubic phases at the expense of the metastable tetragonal phase at the microstructural level. With increasing cubic phase particle size, the substance becomes more porous, resulting in less light scattering at grain boundaries and hence increased translucency [18].

In comparison to the previous two generations of zirconia materials, those cubic zirconia materials exhibit optical qualities that are comparable to those of glass ceramics. However, on a material level, the high translucency obtained by increasing the yttria level reduces the material's intrinsic mechanical properties to a significant degree. As a result, cubic zirconia materials have a flexural strength of around half that of standard 3Y-TZP materials (approximately 600 MPa) [18, 28].

iv. **(Fourth generation 4Y-TZP 0.05** Al_2O_3**)**: as mentioned before 4Y-PSZ zirconia has a translucency and mechanical properties that fall between 3Y-PSZ zirconia and 5Y-PSZ zirconia, making it an appealing material for the esthetic region [26].

v. **Novel Multilayer zirconia**

Construction of zirconia blanks with several, differently-shaded layers and creation of polychromatic zirconia for more esthetic full-contour zirconia restorations than conventional monochromatic zirconia in the form of a gradational multilayered zirconia disc with two to seven or even more layers of color, simulating the appearance of layered porcelain in a full-contour monolithic restorations. Full-contour zirconia restorations with superior esthetics to the monochromatic restorations can be obtained immediately after sintering [15].

This multi-layered technology uses only pigmentation to simulate the shade-gradient of natural teeth while maintaining the same yttria percentage in the zirconia blank, resulting in a color gradient with no difference in flexural intensity between the enamel and dentin layers [23, 36].

What could be called the age of the next generation of modern multi-layered transparent zirconia materials has begun in the last years. These materials (for example, IPS e.max ZirCAD Prime and Katana Multi-Layer) comprise two to four layers of varying translucency inside the same product. This is accomplished by a novel manufacturing procedure that blends two zirconia materials (3Y-TZP and 5Y-TZP) to create a single unique material. By lowering the cubic content of the previous generation to a balanced level with 3Y-TZP, these new materials achieve a balance of strength and esthetics [36, 37].

A multi-layered technology was introduced by combining two generations of zirconia (combination of two different percentages of yttia) in one blank with the aim of combining the benefits of both generations of zirconia. This is mainly a combination of a high-flexural-strength 3Y-TZP in the dentin/body region to enhance flexibility and a high-translucency 5Y-TZP in the incisal or occlusal region to improve esthetics. Even the 5Y-TZP, which has the lowest flexural strength of the zirconia generations, possessed superior mechanical qualities and a translucency comparable to lithium disilicate ceramics [7, 38].

6. Clinical applications of zirconia in dentistry

The construction of veneers, full and partial coverage crowns, implants, fixed partial dentures (FPDs), posts and/or cores, implants, and implant abutments are among the range of zirconia's modern clinical applications. Extra-coronal attachments, surgical drills, cutting burs, and orthodontic brackets are a few additional zirconia-based auxiliary components that are offered as commercial dental products [39].

6.1 Zirconia crown

Space, para-functional habits, malocclusion, short clinical crowns, tooth mobility, tooth inclination considerations, and basic clinical sequence are all the same as they are for other all-ceramic crowns. The tooth preparation clinical guidelines for zirconia crowns are also the same as those for metal-ceramic restorations.

An appropriately constructed diamond set is typically used to accomplish the tooth preparation for a zirconia crown, which should give a desirable distribution of the functional stresses. In general, 1.5 mm to 2.0 mm of incisal or occlusal reduction and 1.2 mm to 1.5 mm of axial reduction are needed to prepare a tooth for a zirconia restoration. All dihedral angles should be tapered, and the axial convergence angle of the crown preparation should be around 6 degrees. The preparation should end with a uniform 0.8 mm to 1.2 mm subgingival (about 0.5 mm) deep chamfer or marginal shoulder with rounded internal angles [39]. **Figure 3** is showing a full coverage of 12 maxillary single zirconia crowns.

6.2 Zirconia fixed partial denture

In comparison to other traditional all-ceramic systems like lithium-disilicate glass ceramics and zirconia-reinforced glass-infiltrated alumina, Y-TZP FPDs were found to have a significantly higher load bearing capacity. It has also been reported that veneering further increased fracture resistance [39, 40]. The most recent framework material for the production of all-ceramic FPDs in either anterior or posterior sites is Y-TZP, which is based on the superior mechanical qualities of zirconia (such as high flexural strength and fracture resistance) [41]. **Figures 4** and **5** are showing zirconia framework in anterior and posterior regions respectively.

6.3 Zirconia as post and core

Meyenberg et al. reported using zirconia in post-and-core systems for the first time in 1995. They found no failure after an average observation period of 11 months [42]. Since then, zirconia posts (ZPs) have emerged as a potential treatment option for teeth with reduced structural integrity and filled roots teeth, particularly for patients with high esthetic needs. Because there are different observation periods, post surface treatments, cement systems, tooth preparation designs, and post designs, the success rate of ZPs in the literature ranges greatly, from 81.3%2 to 100%18. In order to prevent rotation of the posts and cores, for instance, the length of the posts should be prepared to be longer than that of the clinical crowns [43]. Additionally, the structural integrity of the root-filled teeth must have a ferrule with a minimum height of 1.5 mm to counterbalance the lateral forces experienced during post insertion, the wedging effect of posts, and the functional lever forces [44–46]. these restorations also have some problems. For instance, the high elastic modulus may lead to a less uniform stress distribution throughout the tooth, and complications such as tooth fracture and post debonding still exist [47]. **Figure 6** is showing all-zirconia post and core of a maxillary left endodontically treated lateral incisor (tooth 12)) where Laboratory work performed by Mr. F Ferraresso (Saluzzo, Italy) and Dr. SO Koutayas (Corfu, Greece).

6.4 Zirconia implants titanium

Zirconia is emerging as a promising alternative to conventional Titanium based implant system for oral rehabilitation with superior biological, esthetic, mechanical

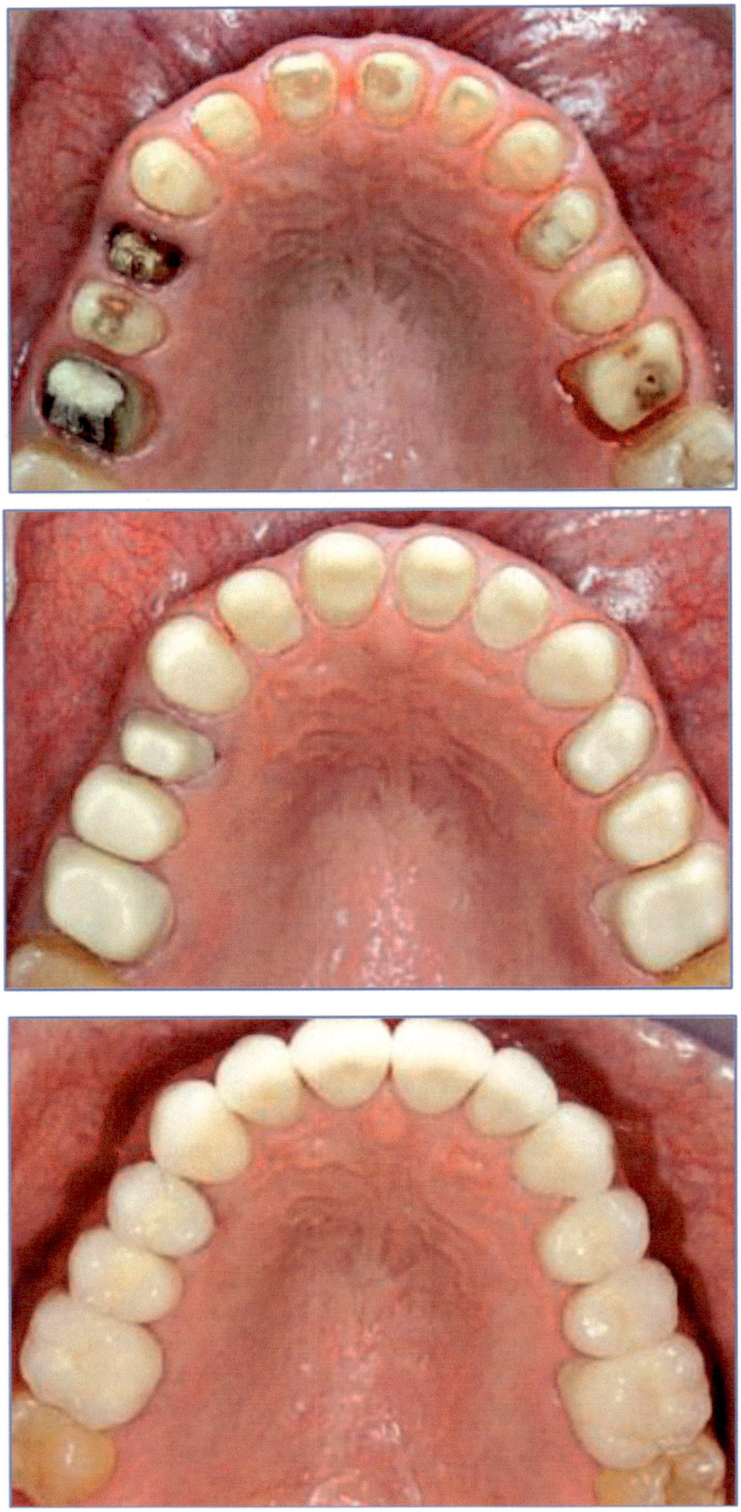

Figure 3.
A total of 12 maxillary single zirconia crowns (teeth 16 to 26). Top: full coverage preparation of the abutment teeth. Middle: Zirconia. Bottom: final clinical situation after crown adhesive cementation [39].

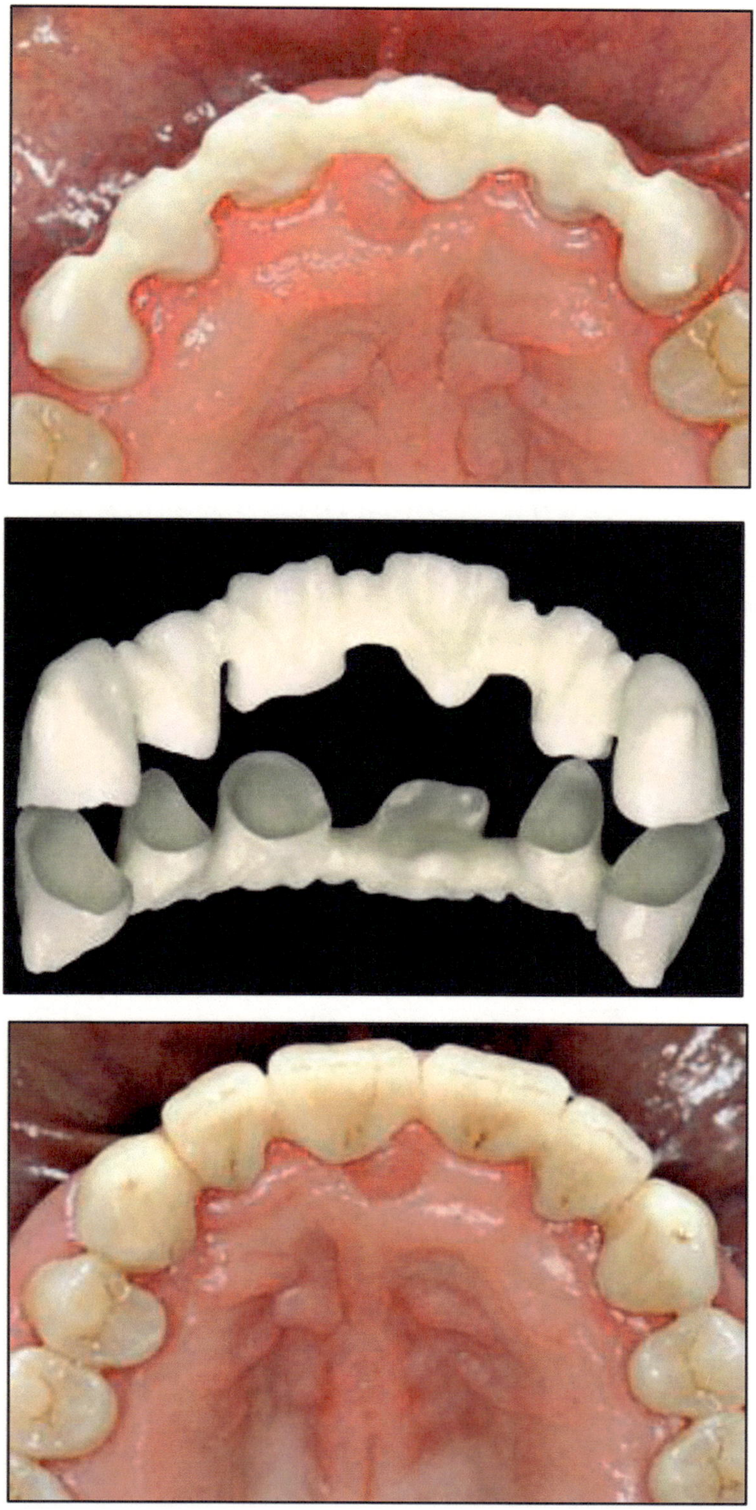

Figure 4.
Anterior six-unit zirconia fixed partial denture restoration (teeth 13 to 23). Top: Zirconia framework in situ. Middle: Zirconia framework after laboratory completion. Bottom: Final clinical situation after cementation [39].

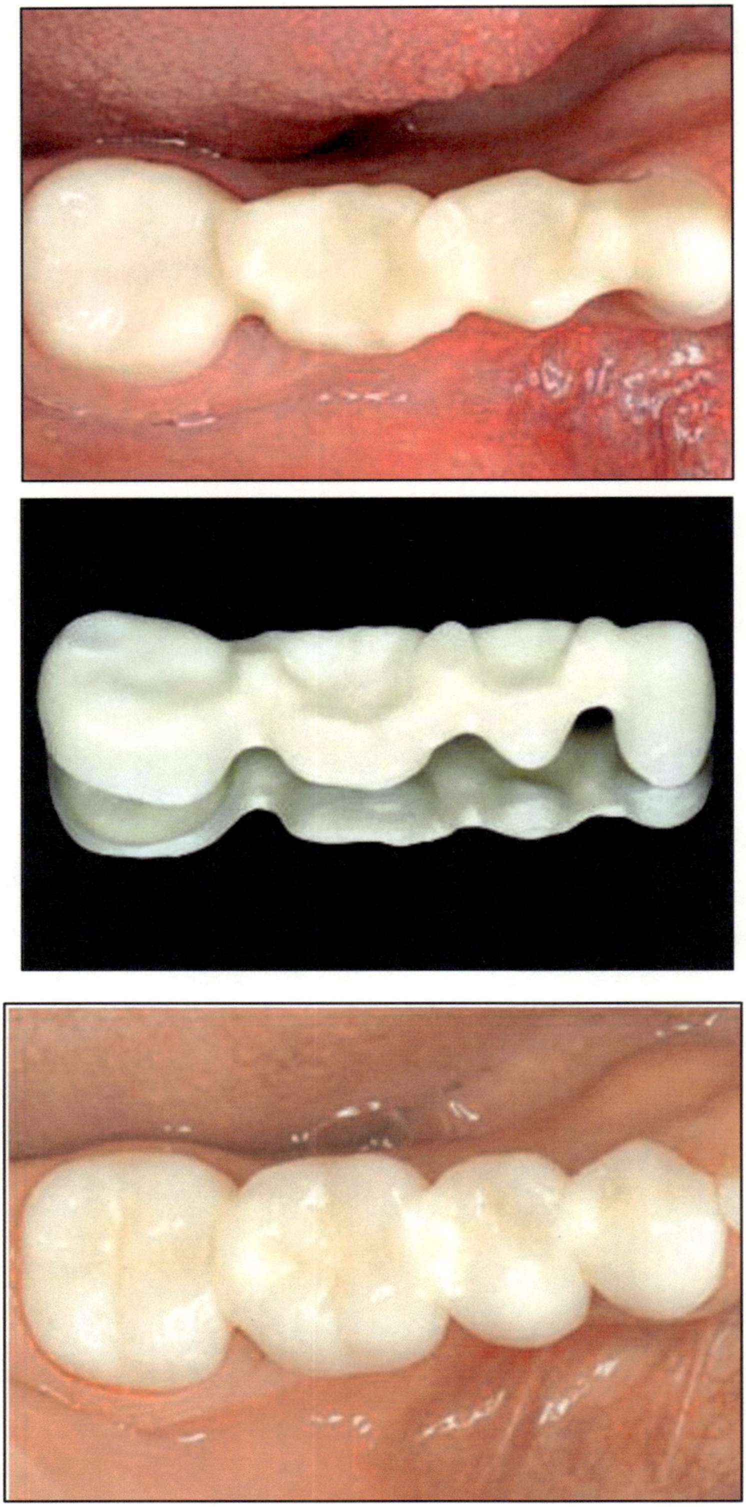

Figure 5.
Posterior four-unit zirconia fixed partial denture restoration (teeth 47 to 44). Top: Zirconia framework. Middle: Zirconia framework after laboratory completion. Bottom: final clinical situation after cementation [39].

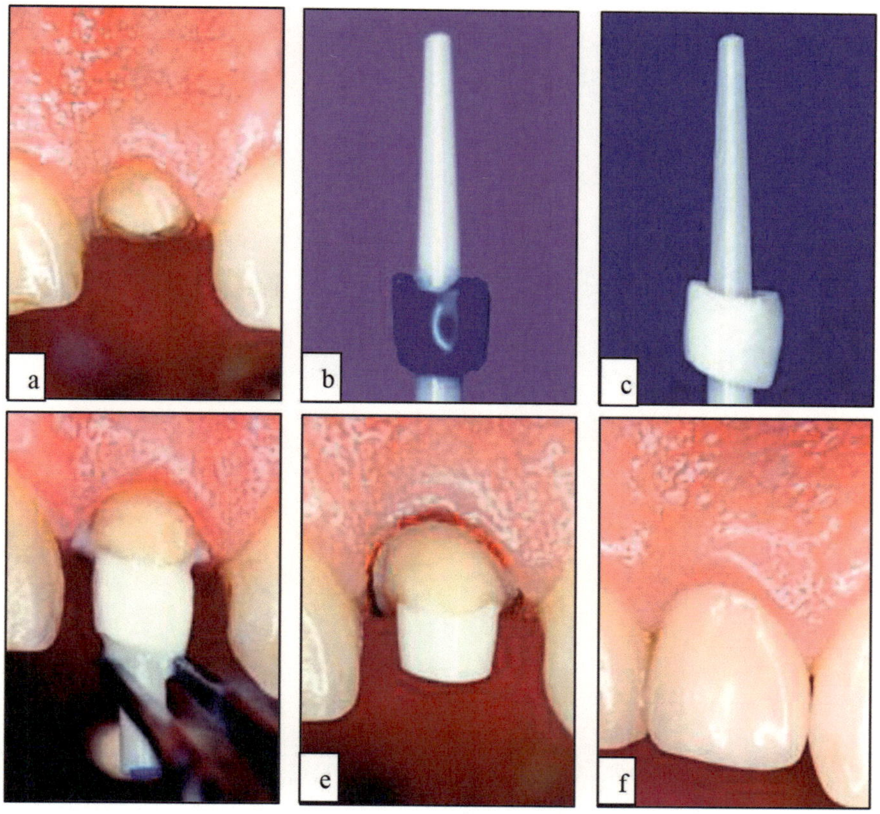

Figure 6.
Single crown restoration of a maxillary left endodontically treated lateral incisor (tooth 12) with the use of an all-zirconia post and core: a) initial situation after endodontic treatment b) prefabricated zirconia post with core analogue model, c) two- piece all-zirconia post and core after milling of a Y-TZP core d) bonding of the post and core restoration using an adhesive resin (Panavia 21, Kuraray), e) completion of the tooth preparation, f) final clinical situation after crown placement [39].

and optical properties [48]. To date, there are five commercially available zirconia implant systems on the market. One- and two-piece designs are offered by only one system (Sigma, Incermed, Lausanne, Switzerland), whereas one-piece designs are only offered by CeraRoot, CeraRoot Dental Implants, Barcelona, Spain; Z- Look3, Z-Systems, Constance, Germany; whiteSKY, Bredent Medical, Senden, Germany; and zit-z, Ziterion, Uffenheim, Germany. A customized zirconia root-analogue implant with a micro- and macro-retentive implant surface was also described in a recent clinical trial, but neither the zirconia material nor the milling machine were fully explained [49]. Zirconia implants do not have any clinical long-term data despite some encouraging preliminary clinical results. 93% of survival rate after one year, according to reports (189 one-piece implants, Z-Systems) [50] 98% (66 one-piece implants, Z-Systems), [51] and 100% (one-piece implants, CeraRoot) [52]. **Figure 7** is showing: Zirconia implant supported zirconia crown (tooth 12) where laboratory work performed by Mr. W Woerner (Freiburg, Germany).

6.5 Zirconia implant abutment

In the most recent systematic review, published in2013, Bidra and Rungruanganunt compared the survival, mechanical, biological, and esthetic outcomes of implant

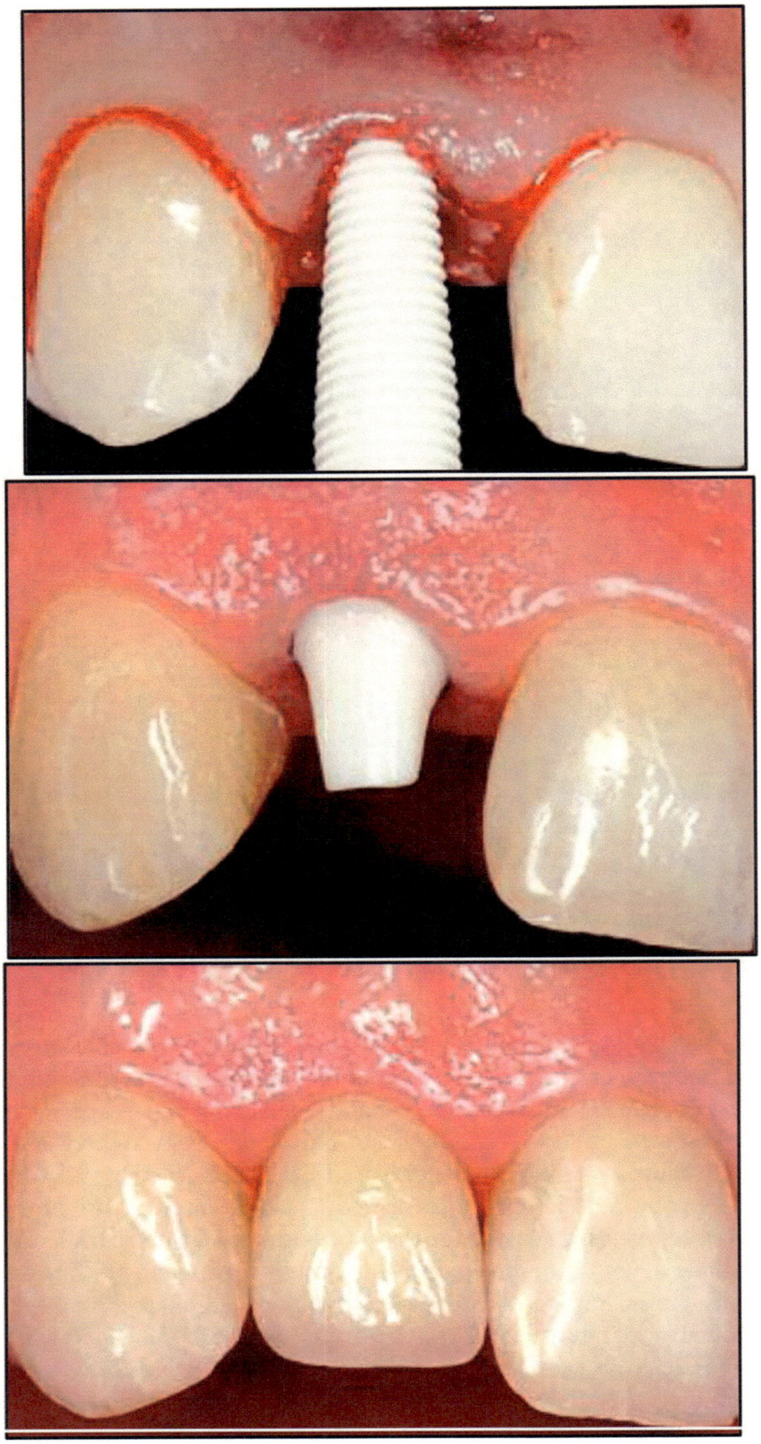

Figure 7.
Zirconia implant supported zirconia crown (tooth 12. Top: Zirconia implant placement after tooth extraction. Middle: 4 months later; placement of retraction cord prior to impression. Bottom: after final cementation of zirconia crown [39].

abutments (Ti and Zir). Due to greater color integration, they came to the conclusion that Zir abutments were preferred from an esthetic standpoint, especially for patients with thin mucosal tissues. They had improved gingival color, according to a recent analysis of their esthetic outcomes, and Zirconia had comparable soft-tissue recession, probing depths, bleeding on probing, marginal bone level, and patient reported outcomes as Ti [53, 54]. Zirconia abutments, however, had greater mechanical issues than Ti abutments. Therefore, the main obstacle to the widespread application of Zir abutments is their lack of mechanical strength [55]. **Figure 8** is showing zirconia prefabricated implant abutment of an upper right lateral incisor (tooth 12) where Clinical and laboratory work performed by Dr. SO Koutayas (Corfu, Greece) and Dr. D Charisis (Athens, Greece), respectively.

6.6 Zirconia dental auxiliary components

different zirconia-based auxiliary components such as cutting burs and surgical drills, extra-coronal attachments, and orthodontic brackets are also available as commercial dental product [39, 56].

6.6.1 Zirconia orthodontic brackets

Orthodontic brackets made of Y-TZP are stronger, more resistant to wear and deformation, less likely to adhere plaque, and more esthetically pleasing. Additionally, they share the same frictional qualities as polycrystalline alumina brackets and show good sliding properties with both stainless steel and nickel-titanium arch wires [56, 57].

6.6.2 Zirconia prefabricated zirconia attachments

The use of Prefabricated zirconia attachments in clinical applications is based on the material's durability and strength. The literature on clinical performance and effectiveness, however, is nonexistent. There are currently two different types of Y-TZP attachments available on the market: an extracoronal, cylindrical, or ball attachment for removably attached partial dentures, and a ball attachment for overdentures that is a part of a zirconia post (Biosnap, Incermed) and is available in three diameters for three levels of retention (Proxisnap, Incermed) [39].

6.6.3 Zirconia cutting and surgical instruments

Newly created zirconia cutting instruments (such as drills and burs) can be employed in soft tissue trimming, maxillofacial surgery, implantology, and other fields. These tools have been shown to be resistant to chemical corrosion, and they provide maximum cutting efficiency while operating smoothly and with less vibration. Lastly, alumina-toughened zirconia (ATZ) can be used to create surgical equipment such as scalpels, tweezers, periosteal elevators, and depth gauges using injection molding (Z- Look3 Instruments, Z-Systems) [39].

7. Conclusions

It is obvious that zirconia is continuous developing and growing ceramic material for wide applications in dentistry from the traditional core or framework material

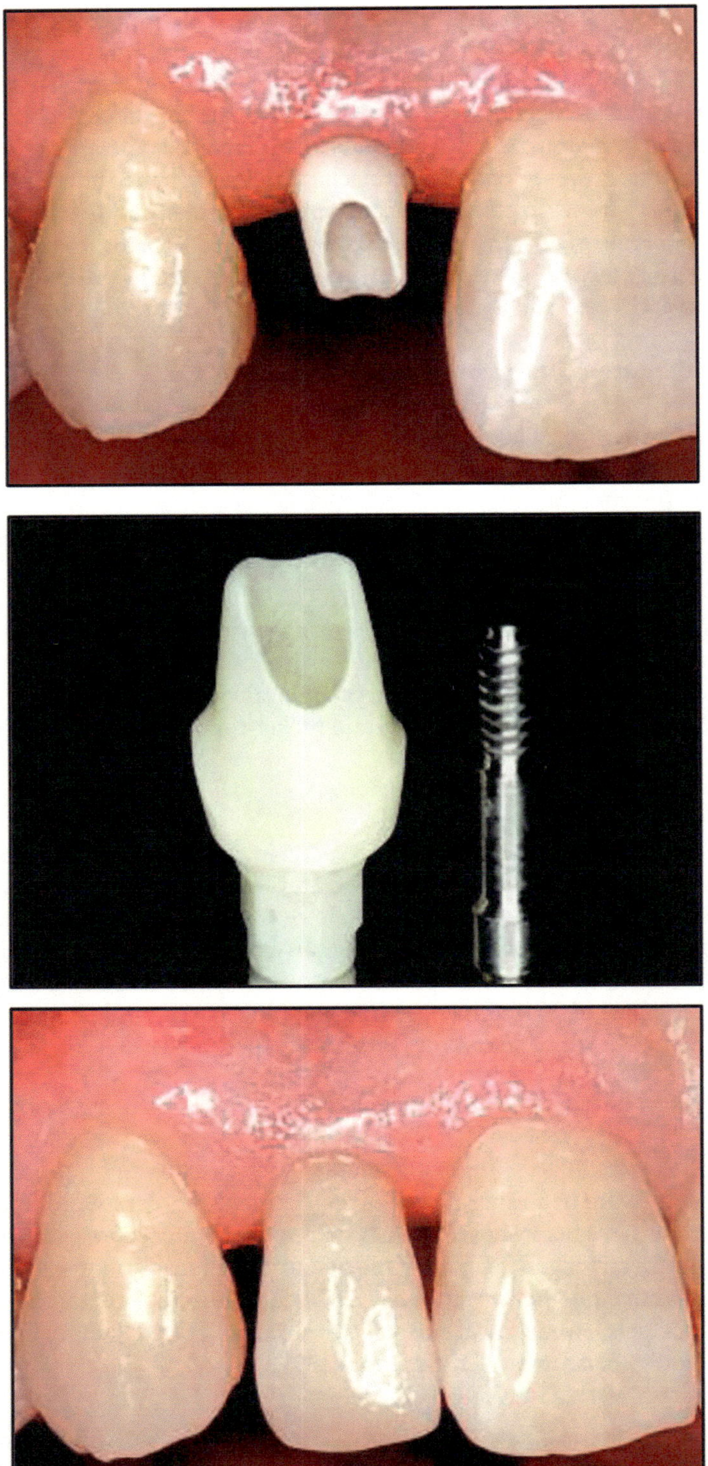

Figure 8.
Single implant all-ceramic crown restoration with the use of a zirconia prefabricated of an upper right lateral incisor (tooth 12). Top: abutment connection. Middle: Zirconia abutment after laboratory modification and Ti screw. Bottom: final clinical situation after crown adhesive cementation [39].

to the newly developed monolithic multilayer zirconia Dental zirconia continues to increase and is classified into many species in the yttria system alone. They are classified with yttria content, monochromatic/polychromatic, uniform/hybrid composition, and monolayer/multilayer. Zirconia with a higher yttria content is more translucent and less strong mechanically. Zirconia applications seem to consolidate a well-established position in clinical dentistry, due to the improvements in CAD/CAM technology and to the material's exceptional physical properties. Existing clinical studies demonstrated a promising survival potential regarding tooth-supported restorations therefore, a suitable zirconia should be selected depending on whether strength or esthetics are desired. Therefore, it is concluded that an adequately selected zirconia is a suitable material because of its mechanically, esthetically, and biologically excellent properties. Zirconia abutments provide a favorable bioesthetic addition to implant dentistry, however, long-term clinical assessment is needed for accurate evaluation of implant-supported zirconia restorations.

Acknowledgements

Great thanks to the college of dentistry/university of Mosul/Iraq for their continuous support and encouragement.

Conflict of interest

The authors declare no conflict of interest.

Abbreviations

3Y-HA	conventional TZP stabilized with 3-mol% yttria and added with a relatively high content of alumina (0.25–0.5 wt%)
3Y	high translucent TZP stabilized with 3-mol% yttria and added with a relatively low content of alumina (less than 0.05 wt%)
4Y	high strength PSZ stabilized with 4-mol% yttria and added with a relatively low content of alumina (less than 0.05 wt%)
5Y	high translucent PSZ stabilized with 5-mol% yttria and added with a relatively low content of alumina (less than 0.05 wt%)
6Y	super high translucent PSZ stabilized with 6-mol% yttria and added with a relatively low content of alumina (less than 0.05 wt%)
M3Y	polychromatic multilayered 3Y
M4Y	polychromatic multilayered 4Y
M5Y	polychromatic multilayered 5Y
M6Y	polychromatic multilayered 6Y
M3Y-5Y	polychromatic multilayer with hybrid composition from 3Y to 5Y
M3Y-4Y	polychromatic multilayer with hybrid composition from 3Y to 4Y
M4Y-5Y	polychromatic multilayer with hybrid composition from 4Y to 5Y

Author details

Ali Dahee Malallah* and Nadia Hameed Hasan
Department of Conservative Dentistry, College of Dentistry, University of Mosul, Mosul, Iraq

*Address all correspondence to: ali.dhahi@uomosul.edu.iq

References

[1] Denry I, Kelly JR. State of the art of zirconia for dental applications. Dental Materials. 2008;**24**(3):299-307. DOI: 10.1016/j.dental.2007.05.007

[2] Sehgal M, Bhargava A, Gupta S, Gupta P. Shear bond strengths between three different Yttria-stabilized zirconia dental materials and veneering ceramic and their susceptibility to autoclave induced low-temperature degradation. International Journal of Biomaterials. 2016;**2016**:9658689. DOI: 10.1155/2016/9658689

[3] Nisticò R. Zirconium oxide and the crystallinity hallows. Journal of the Australian Ceramic Society. 2021;**57**(1):225-236

[4] Manicone PF, Rossi Iommetti P, Raffaelli L. An overview of zirconia ceramics: Basic properties and clinical applications. Journal of Dentistry. 2007;**35**(11):819-826. DOI: 10.1016/j.jdent.2007.07.008

[5] Saker S, Özcan M. Marginal discrepancy and load to fracture of monolithic zirconia laminate veneers: The effect of preparation design and sintering protocol. Dental Materials Journal. 2021;**40**(2):331-338. DOI: 10.4012/dmj.2020-007

[6] Oba Y, Koizumi H, Nakayama D, Ishii T, Akazawa N, Matsumura H. Effect of silane and phosphate primers on the adhesive performance of a tri-n-butylborane initiated luting agent bonded to zirconia. Dental Materials Journal. 2014;**33**(2):226-232. DOI: 10.4012/dmj.2013-346

[7] Preis V, Behr M, Hahnel S, Handel G, Rosentritt M. In vitro failure and fracture resistance of veneered and full-contour zirconia restorations. Journal of Dentistry. 2012;**40**(11):921-928. DOI: 10.1016/j.jdent.2012.07.010

[8] Michailova M, Elsayed A, Fabel G, Edelhoff D, Zylla IM, Stawarczyk B. Comparison between novel strength-gradient and color-gradient multilayered zirconia using conventional and high-speed sintering. Journal of the Mechanical Behavior of Biomedical Materials. 2020;**111**:103977. DOI: 10.1016/j.jmbbm.2020.103977

[9] Zarone F, Di Mauro MI, Ausiello P, Ruggiero G, Sorrentino R. Current status on lithium disilicate and zirconia: A narrative review. BMC Oral Health. 2019;**19**(1):134. DOI: 10.1186/s12903-019-0838-x

[10] Ban S. Classification and properties of dental zirconia as implant fixtures and superstructures. Materials (Basel). 2021;**14**(17):4879. Published 2021 Aug 27. DOI: 10.3390/ma14174879

[11] Furtado de Mendonca A, Shahmoradi M, Gouvêa CVD, De Souza GM, Ellakwa A. Microstructural and mechanical characterization of CAD/CAM materials for monolithic dental restorations. Journal of Prosthodontics. 2019;**28**(2):e587-e594. DOI: 10.1111/jopr.12964

[12] Chevalier J, Gremillard L, Virkar AV, Clarke DR. The tetragonal monoclinic transformation in zirconia: Lessons learned and future trends. Journal of the American Ceramic Society. 2009;**92**(9):1901-1920. DOI: org/10.1111/j.1551-2916.2009.03278.x

[13] Nakamura K, Harada A, Kanno T, et al. The influence of low-temperature degradation and cyclic loading on

the fracture resistance of monolithic zirconia molar crowns. Journal of the Mechanical Behavior of Biomedical Materials. 2015;**47**:49-56. DOI: 10.1016/j.jmbbm.2015.03.007

[14] Silva LHD, Lima E, Miranda RBP, Favero SS, Lohbauer U, Cesar PF. Dental ceramics: A review of new materials and processing methods. Brazilian Oral Research. 2017;**31**(suppl. 1):e58. Published 2017 Aug 28. DOI: 10.1590/1807-3107BOR-2017.vol31.0058

[15] Sulaiman TA, Abdulmajeed AA, Donovan TE, Vallittu PK, Närhi TO, Lassila LV. The effect of staining and vacuum sintering on optical and mechanical properties of partially and fully stabilized monolithic zirconia. Dental Materials Journal. 2015;**34**(5):605-610. DOI: 10.4012/dmj.2015-054

[16] Pereira GKR, Guilardi LF, Dapieve KS, Kleverlaan CJ, Rippe MP, Valandro LF. Mechanical reliability, fatigue strength and survival analysis of new polycrystalline translucent zirconia ceramics for monolithic restorations. Journal of the Mechanical Behavior of Biomedical Materials. 2018;**85**:57-65. DOI: 10.1016/j.jmbbm.2018.05.029

[17] Hatanaka GR, Polli GS, Adabo GL. The mechanical behavior of high-translucent monolithic zirconia after adjustment and finishing procedures and artificial aging. The Journal of Prosthetic Dentistry. 2020;**123**(2):330-337. DOI: 10.1016/j.prosdent.2018.12.013

[18] Stawarczyk B, Keul C, Eichberger M, Figge D, Edelhoff D, Lümkemann N. Three generations of zirconia: From veneered to monolithic. Part I. Quintessence International. 2017;**48**(5):369-380. DOI: 10.3290/j.qi.a38057

[19] Flinn BD, Raigrodski AJ, Mancl LA, Toivola R, Kuykendall T. Influence of aging on flexural strength of translucent zirconia for monolithic restorations. The Journal of Prosthetic Dentistry. 2017;**117**(2):303-309. DOI: 10.1016/j.prosdent.2016.06.010

[20] Elsaka SE. Optical and mechanical properties of newly developed monolithic multilayer zirconia. Journal of Prosthodontics. 2019;**28**(1):e279-e284. DOI: 10.1111/jopr.12730

[21] Ueda K, Güth JF, Erdelt K, Stimmelmayr M, Kappert H, Beuer F. Light transmittance by a multi-coloured zirconia material. Dental Materials Journal. 2015;**34**(3):310-314. DOI: 10.4012/dmj.2014-238

[22] Alsadon O, Patrick D, Johnson A, Pollington S, Wood D. Fracture resistance of zirconia-composite veneered crowns in comparison with zirconia-porcelain crowns. Dental Materials Journal. 2017;**36**(3):289-295. DOI: 10.4012/dmj.2016-298

[23] Kaizer MR, Kolakarnprasert N, Rodrigues C, Chai H, Zhang Y. Probing the interfacial strength of novel multi-layer zirconias. Dental Materials. 2020;**36**(1):60-67. DOI: 10.1016/j.dental.2019.10.008

[24] El-Ghany OSA, Sherief AH. Zirconia based ceramics, some clinical and biological aspects: Review, future. Dental Journal. 2016;**2**(2):55-64. ISSN 2314-7180. DOI: 10.1016/j.fdj.2016.10.002

[25] Grech J, Antunes E. Zirconia in dental prosthetics: A literature review. Journal of Materials Research and Technology. 2019;**8**(5):4956-4964, ISSN 2238-7854. DOI: 10.1016/j.jmrt.2019.06.043

[26] Zhang Y, Lawn BR. Novel zirconia materials in dentistry. Journal of Dental

Research. 2018;**97**(2):140-147. DOI: 10.1177/0022034517737483

[27] Abdulmajeed A, Sulaiman T, Abdulmajeed A, Bencharit S, Närhi T. Fracture load of different zirconia types: A mastication simulation study. Journal of Prosthodontics. 2020;**29**(9):787-791. DOI: 10.1111/jopr.13242

[28] Kwon SJ, Lawson NC, McLaren EE, Nejat AH, Burgess JO. Comparison of the mechanical properties of translucent zirconia and lithium disilicate. The Journal of Prosthetic Dentistry. 2018;**120**(1):132-137. DOI: 10.1016/j.prosdent.2017.08.004

[29] Burgess JO. Zirconia: The material, its evolution, and composition. The Compendium of Continuing Education in Dentistry. 2018;**39**(suppl. 4):4-8

[30] Fabris S, Paxton AT, Finnis MW. A stabilization mechanism of zirconia based on oxygen vacancies only. Acta Materialia. 2002;**50**(20):5171-5178,ISSN 1359-6454. DOI: 10.1016/S1359-6454(02)00385-3

[31] Kara R. Effects of different sintering times on the adaptation of monolithic zirconia crowns. J Health Med Sc. 2020;**3**(4):449-456. DOI: 10.31014/aior.1994.03.04.139

[32] Jiang L, Liao Y, Wan Q, Li W. Effects of sintering temperature and particle size on the translucency of zirconium dioxide dental ceramic. Journal of Materials Science. Materials in Medicine. 2011;**22**(11):2429-2435. DOI: 10.1007/s10856-011-4438-9

[33] Alraheam IA, Donovan T, Boushell L, Cook R, Ritter AV, Sulaiman TA. Fracture load of two thicknesses of different zirconia types after fatiguing and thermocycling. The Journal of Prosthetic Dentistry. 2020;**123**(4):635-640. DOI: 10.1016/j.prosdent.2019.05.012

[34] Guazzato M, Albakry M, Quach L, Swain MV. Influence of grinding, sandblasting, polishing and heat treatment on the flexural strength of a glass-infiltrated alumina-reinforced dental ceramic. Biomaterials. 2004;**25**(11):2153-2160. DOI: 10.1016/j.biomaterials.2003.08.056

[35] Daou EE. The zirconia ceramic: Strengths and weaknesses. Open. Dental Journal. 2014;**8**:33-42. Published 2014 Apr 18. DOI: 10.2174/18742106014 08010033

[36] Kolakarnprasert N, Kaizer MR, Kim DK, Zhang Y. New multi-layered zirconias: Composition, microstructure and translucency. Dental Materials. 2019;**35**(5):797-806. DOI: 10.1016/j.dental.2019.02.017

[37] Mao L, Kaizer MR, Zhao M, Guo B, Song YF, Zhang Y. Graded ultra-translucent zirconia (5Y-PSZ) for strength and functionalities. Journal of Dental Research. 2018;**97**(11):1222-1228. DOI: 10.1177/0022034518771287

[38] Lawson NC, Maharishi A. Strength and translucency of zirconia after high-speed sintering. Journal of Esthetic and Restorative Dentistry. 2020;**32**(2):219-225. DOI: 10.1111/jerd.12524

[39] Luthy H, Filser F, Loeffel O, Schumacher M, Gauckler LJ, Hammerle CH. Strength and reliability of four-unit all-ceramic posterior bridges. Dental Materials. 2005;**21**:930-937

[40] Tinschert J, Natt G, Mautsch W, Augthun M, Spiekermann H. Fracture resistance of lithium disilicate-, alumina-, and zirconia-based three-unit fixed partial dentures: A laboratory study. The International Journal of Prosthodontics. 2001;**14**:231-238

[41] Koutayas SO, Vagkopoulou T, Pelekanos S, Koidis P, Strub JR. Zirconia

in dentistry: Part 2. Evidence-based clinical breakthrough. Eur. Journal of Esthetic Dentistry. 2009;**4**(4):348-380

[42] Meyenberg KH, Lüthy H, Schärer P. Zirconia posts: A new all-ceramic concept for nonvital abutment teeth. Journal of Esthetic Dentistry. 1995;**7**:73-80

[43] Nothdurft FP, Pospiech PR. Clinical evaluation of pulpless teeth restored with conventionally cemented zirconia posts: A pilot study. The Journal of Prosthetic Dentistry. 2006;**95**:311-314

[44] Standlee J, Caputo A. Interaction of endodontic posts with tooth structure. In: Kurer P, editor. Kurer Anchor System. Chicago: Quintessence; 1984. pp. 140-160

[45] Standlee JP, Caputo AA, Collard EW, Pollack MH. Analysis of stress distribution by endodontic posts. Oral Surgery, Oral Medicine, and Oral Pathology. 1972;**33**:952-960

[46] Caputo AA, Standlee JP, editors. Prerestorative endodontics. In: Biomechanics in Clinical Dentistry. Chicago: Quintessence; 1987. pp. 85-96

[47] Zhang X, Pei X, Pei X, Wan Q, Chen J, Wang J. Success and complication rates of root-filled teeth restored with zirconia posts: A critical review. The International Journal of Prosthodontics. 2019;**32**(5):411-419. DOI: 10.11607/ijp.6179

[48] Sivaraman K, Chopra A, Narayan AI, Balakrishnan D. Is zirconia a viable alternative to titanium for oral implant? A critical review. Journal of Prosthodontic Research. 2018;**62**(2):121-133. DOI: 10.1016/j.jpor.2017.07.003

[49] Pirker W, Kocher A. Immediate, non-submerged, root- analogue zirconia implant in single tooth replacement. International Journal of Oral and Maxillofacial Surgery. 2008;**37**:293-295

[50] Mellinghoff J. First clinical results of dental screw implants made of zirconium oxide. Zahnärztl Impl. 2006;**22**:288-293

[51] Volz U, Blaschke C. Metal-free reconstruction with zirconia implants and zirconia crowns. Quintessence Journal of Dental Technology. 2004;**2**:324-330

[52] Oliva J, Oliva X, Oliva JD. One-year follow-up of first consecutive 100 zirconia dental implants in humans: A comparison of 2 different rough surfaces. The International Journal of Oral & Maxillofacial Implants. 2007;**22**:430-435

[53] Bidra AS, Rungruanganunt P. Clinical outcomes of implant abutments in the anterior region: A systematic review. Journal of Esthetic and Restorative Dentistry. 2013;**25**:159-176

[54] Linkevicius T, Vaitelis J. The effect of zirconia or titanium as abutment material on soft peri-implant tissues: A systematic review and meta-analysis. Clinical Oral Implants Research. 2015;**26**(Suppl. 11):139-147

[55] Naveau A, Rignon-Bret C, Wulfman C. Zirconia abutments in the anterior region: A systematic review of mechanical and esthetic outcomes. The Journal of Prosthetic Dentistry. 2019;**121**(5):775-781.e1. DOI: 10.1016/j.prosdent.2018.08.005

[56] Kusy RP. Orthodontic biomaterials: From the past to the present. The Angle Orthodontist. 2002;**72**:501-512

[57] Keith O, Kusy RP, Whitley JQ. Zirconia brackets: An evaluation of morphology and coefficients of friction. American Journal of Orthodontics and Dentofacial Orthopedics. 1994;**106**:605-614

Section 3

Production of Zirconia

Chapter 3

Zirconia: Synthesis and Characterization

Bincy Cyriac

Abstract

Main resource of zirconia is the mineral zircon which occurs in beach sand and placer deposits. Alkali fusion and thermal plasma dissociation are the frequently adopted procedures to convert zircon to zirconia. Synthesis of different zirconia phases (monoclinic, cubic, and tetragonal) can be accomplished by the precise control of different operating parameters and stoichiometry of the reagents. Mesoporous and nano-zirconia which find wide application in catalysis and electronics are synthesized by different methods like solution combustion synthesis, sol–gel synthesis, hydrothermal synthesis, co-precipitation, and solid-phase sintering. Recently, biosynthesis of zirconia has taken a quantum leap due to environmental concerns. The synthesized zirconia is characterized by various chemical, physical, and instrumental methods to find out composition, crystal structure, size, and morphology.

Keywords: zircon, zirconia, plasma dissociated zircon, synthesis of zirconia, characterization of zirconia

1. Introduction

Zirconia (ZrO_2) is one of the important materials which finds wide usage depending on its purity and crystal structure in various fields such as ceramics, refractories, electronics, and others. Diversified applications of zirconia as a high-technology material for industrial applications are due to its superior mechanical, thermal, electrical, chemical, and optical properties [1]. It is an ideal material for the production of ceramics, electronic materials, and pigment due to the combination of properties like hardness, strength, high melting point, and biocompatibility. Zirconia ceramics have excellent biocompatibility with the human body, a property which helped it to replace alumina for prosthesis devices in hip joints, femoral ball beads, and dental implants. Optically clear cubic zirconia, known as synthetic diamond, is widely used in jewelery.

Zirconium oxide exhibits three well-defined crystal structures, i.e., monoclinic, tetragonal, and cubic (**Figure 1**) [2]. The monoclinic phase is stable up to 1170°C, and above this temperature it is transformed into tetragonal phase. The tetragonal phase is stable up to 2370°C and then transforms to the cubic phase, which is stable up to the melting temperature of 2680°C. On cooling to the transformation temperature, the structure reverts back to the original phase [3]. Out of these, tetragonal to monoclinic

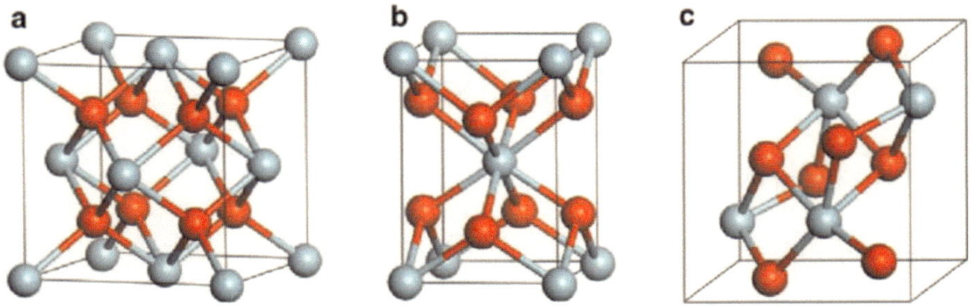

Figure 1.
Crystal structures of ZrO2: (a) cubic, (b) tetragonal, and (c) monoclinic. Red and blue spheres correspond to oxygen and zirconium atoms, respectively.

transformation is of great importance due to large volume change. This volume reduction is very advantageous for improving strength and toughness of ceramics.

Baddeleyite (ZrO_2) is the naturally occurring zirconia, and it occurs in carbonatite rocks [4]. However, the major source of zirconia is its silicate mineral zircon ($ZrSiO_4$), and it mainly occurs as a constituent of beach sand and placer deposits along with rutile, ilmenite, and monazite. Beach sands of Australia, India, Brazil, and USA are rich in zircon. In terms of zirconia, world reserves of zirconium are around 64 million tones. Australia (35%) is the major producer of zircon followed by South Africa (28%), USA, and Mozambique (7% each) [5]. In placer deposits, zircon occurs as a major constituent along with ilmenite, rutile, and quartz, while the minor constituents are sillimanite, garnet, and magnetite. Zircon is separated from beach sands and placer deposits by physical beneficiation method as given in **Figure 2**.

Zircon has the general composition of 67% zirconia and 32.8% silica. Zircon is zirconium hafnium silicate mineral with general formula (Zr, Hf)SiO_4. Zircon usually contains some hafnium, typically about 1%. **Table 1** gives the chemical composition of Egyptian zircon.

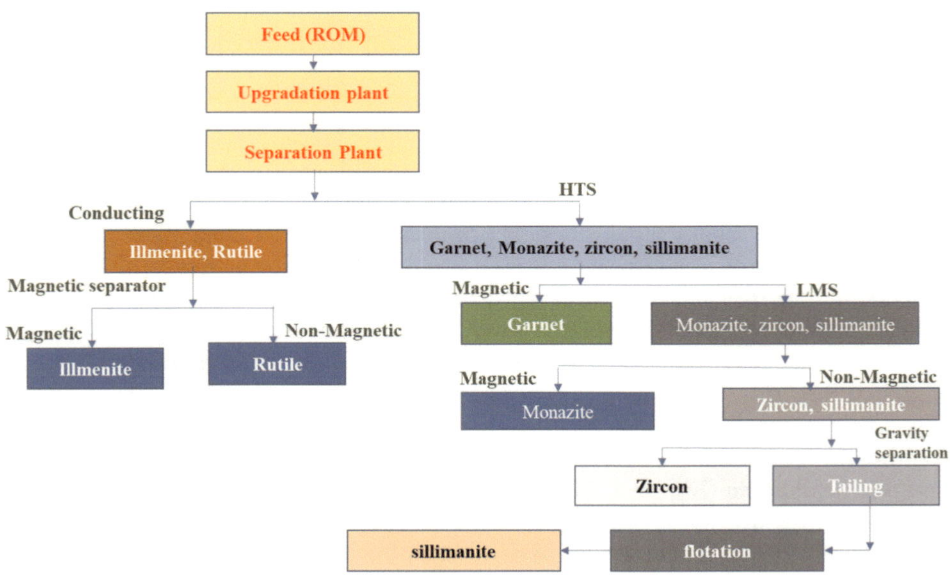

Figure 2.
Physical beneficiation of beach sand [6].

Oxide	% Composition
ZrO_2	61.9
SiO_2	30.6
Fe_2O_3	3.3
HfO_2	1.2
ThO_2	0.56
Na_2O	0.12
MgO	0.18
Al_2O_3	0.87
Cr_2O_3	0.043
U_3O_8	0.039

Table 1.
XRF analysis data of chemical composition of Egyptian zircon [8].

Zircon is chemically very stable. Zircon is considered a refractive material due to its low coefficient of thermal expansion and high melting point. Extraction of zirconium, zirconia, and other products from zircon requires rigorous chemical and thermal treatments to break the bonds between ZrO_2 and SiO_2 (**Figure 3**) [1]. A variety of techniques have been proposed for the extraction of zirconia from zircon.

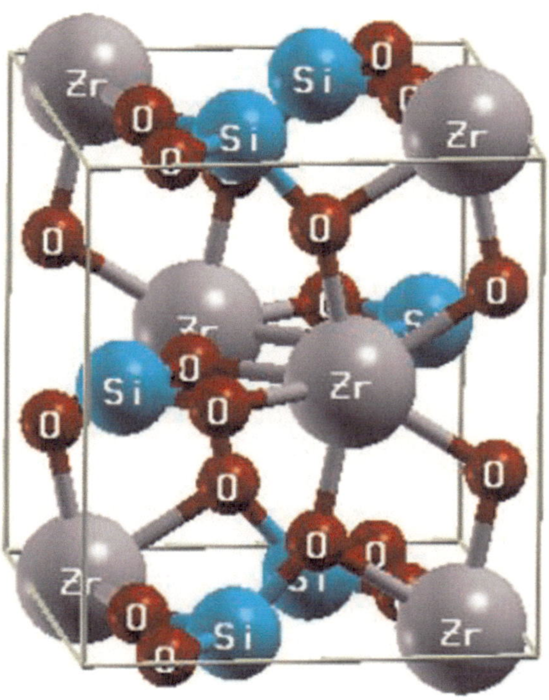

Figure 3.
Crystal structure of zircon [7].

2. Decomposition of zircon

The two major routes used for the decomposition of zircon are chemical and thermal process.

2.1 Chemical process

The chemical decomposition is divided into different categories based on the reagents used as fluxes. They are fused with:

1. Sodium hydroxide (caustic soda) [6, 8–11].

2. Sodium carbonate [12].

3. Calcium oxide and magnesium oxide [13].

4. Calcium carbonate (lime) [14].

5. Carbochlorination [15].

2.1.1 Fusion with sodium hydroxide

This is the widely adopted commercial method for the extraction of zirconia from zircon sand. In this method, zircon is fused at 600°C with caustic soda to produce sodium zirconate (Na_2ZrO_3) and sodium silicate (Na_2SiO_3) (eq. 1):

$$ZrSiO_4 + 4NaOH \rightarrow Na_2ZrO_3\downarrow + Na_2SiO_3 + 2H_2O \quad (1)$$

Water-soluble sodium silicate is removed by washing, and the residual sodium zirconate is treated with hydrochloric acid to produce zirconyl oxychloride ($ZrOCl_2$-ZOC) (eq. 2):

$$Na_2ZrO_3 + 4HCl \rightarrow ZrOCl_2 + NaCl + 2H_2O. \quad (2)$$

Zirconium oxychloride obtained is further acidified to precipitate it as zirconium oxychloride crystals:

$$ZrOCl_{2(l)} + 8H_2O + HCl \rightarrow ZrOCl_2.8H_2O(s) + HCl. \quad (3)$$

The residual solids are separated by filtration which contains mainly sodium zirconate, hydrous zirconia, and some silica. This residue is dissolved in 5 M HCl at 90°C to obtain a clear solution of zirconium oxychloride. The zirconium oxychloride obtained is neutralized with ammonia to precipitate zirconium hydroxide which in turn on calcination at 900°C gives pure zirconia, assaying >99.5%, with a monoclinic structure.

Flow sheet of alkali fusion of zircon is given in **Figure 4**.

2.1.2 Fusion with sodium carbonate

In this method, zircon is fused with sodium carbonate at 1100°C for several hours in an electric furnace to produce sodium zirconium silicate (Na_2ZrSiO_5):

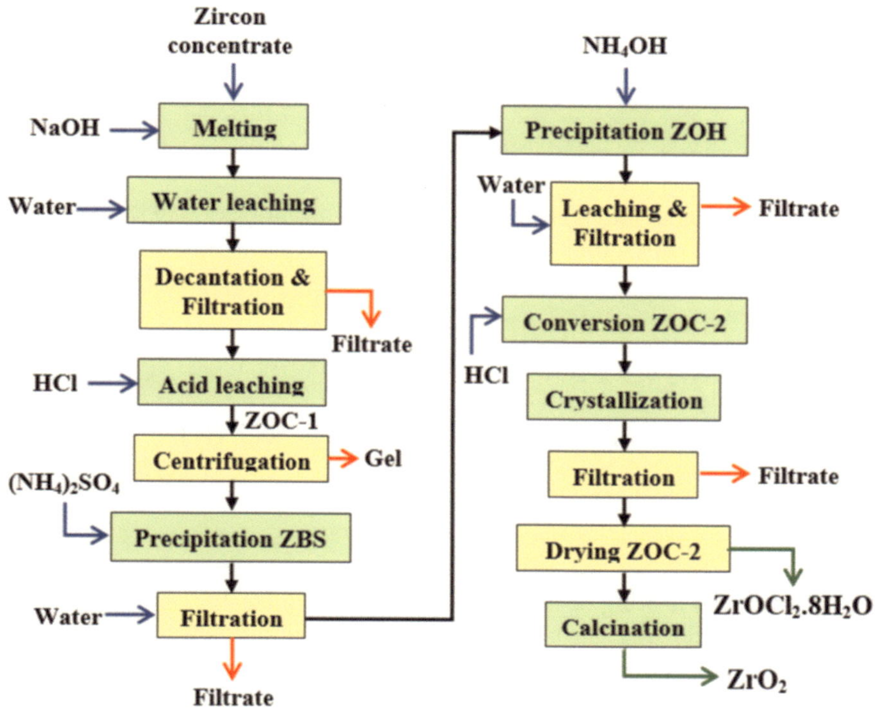

Figure 4.
Flow sheet for alkali fusion of zircon [16].

$$ZrSiO_4 + Na_2CO_3 \rightarrow Na_2ZrSiO_5 + CO_2. \tag{4}$$

Sodium zirconium silicate reacts with excess of sodium carbonate to form sodium zirconate and sodium silicate according to the eqn:

$$Na_2ZrSiO_5 + Na_2CO_3 \rightarrow Na_2ZrO_3 + Na_2SiO_3 + CO_2. \tag{5}$$

Sodium zirconate remaining after the washing is treated with HCl to produce zirconium oxychloride crystal (ZOC) which in turn is converted into zirconia.

2.1.3 Fusion with calcium oxide and magnesium oxide

This method is mainly adopted for the preparation of cubic zirconia. Mixtures of zircon and CaO/MgO in the same molar ratio are fused at 1200°C. The complete disintegration of zircon produces zirconia (ZrO_2) and calcium magnesium silicate according to the equation:

$$ZrSiO_4 + CaO + MgO \rightarrow ZrO_2 + CaMgSiO_4\downarrow. \tag{6}$$

Leaching of the fused mass with hydrochloric acid removes the silica totally. The zirconium oxychloride obtained upon calcination gives cubic zirconia. The temperature and the composition of CaO and MgO play an important role in the concentration of cubic and monoclinic zirconia obtained. At lower percentage of CaO/MgO mixture, monoclinic zirconia content increases whereas at higher CaO/MgO percentage cubic zirconia increases. The fusion of zircon with CaO and MgO separately results in the formation of monoclinic zircon, while their combination in the same molar ratio as zircon produces cubic zircon. At a lower temperature of 1200°C monoclinic zirconia content increases, while at 1400–1500°C cubic zirconia content increases.

2.1.4 Fusion with calcium carbonate (lime)

This method is not an industrial method for the decomposition of zircon. Zircon on fusion with lime and subsequent cooling, produce a very fine powder of calcium silicate and coarse crystals of calcium zirconate enabling the physical separation of the two an easy process:

$$2ZrSiO_4 + 5CaCO_3 \rightarrow 2CaZrO_3 + (CaO)_3(SiO_2)_2 + CO_2. \qquad (7)$$

Calcium zirconate is converted into ZOC by treating with HCl.

2.1.5 Carbochlorination

This reaction takes place in a chlorinator. Zircon is heated with chlorine gas at 1100°C by induction heating of the graphite walls of chlorinator. Products, consisting of zirconium tetrachloride, silicon tetrachloride, and carbon monoxide, which are in gas phase are cooled down to 200°C:

$$ZrSiO_4 + 4Cl + 4C \rightarrow ZrCl_4 + SiC_4 + CO. \qquad (8)$$

On cooling, zirconium tetrachloride solidifies first followed by silicon tetrachloride facilitating their separation. Zirconium tetrachloride is converted into zirconium oxychloride on treatment with water.

Zircon decomposition by alkali fusion has the capability of large-scale production and high efficiency, in comparison with chloride and lime sintering method. But in the former method the temperature profile and atmospheric control inside the furnace plays a crucial role. Silica carryover to zirconium fraction increases at high temperature. Once the sodium silicate is extracted from the sodium zirconate and dissolved in hydrochloric acid, two distinct routes can be followed to precipitate various zirconium chemicals. The most common route is to precipitate zirconium oxychloride crystals (ZOC), with subsequent purification from all contaminants (crystal route). Less known is the process (liquid route) that involves the direct precipitation of zirconium basic sulfate (ZBS). This route will yield a less pure product, with contaminants such as silica and titanium. An important factor in this route is the prevention of silica gel formation, which could hamper final product filtration.

2.2 Thermal decomposition of zircon

Even though chemical decomposition by alkali fusion at very high temperature is the time-tested method for the extraction of zirconia from zircon, it has certain shortcomings. High treatment cost, complexity of the process, and environmental issues associated with effluent treatment are some of them. These drawbacks compelled the search for cost-effective and environmentally friendly alternatives for zircon decomposition. The promising thermal plasma technology offers a one-step alternative for this conversion. Use of thermal plasma dissociation for the decomposition of zircon was carried out first by Wilks and co-workers in 1966 [17]. Zircon is dissociated to ZrO_2 and SiO_2 when heated above 1700°C [18, 19] as expressed by the following reaction:

$$ZrSiO_4 \rightarrow ZrO_2 + SiO_2. \tag{9}$$

The above reaction is reversible and the oxides recombine to form the silicate on cooling. However, if zircon is heated to temperatures exceeding 1900°K and the reaction products are quenched rapidly, the reversible reaction is prevented. Recombination of products back to zircon is prevented by rapid cooling. The silica can be separated from the product by simple acid or alkali leaching to get pure zirconium oxide [20]. For thermal plasma decomposition, a direct current non-transferred argon arc plasma torch is used to generate plasma [21, 22]. The thermal dissociation is thermodynamically reversible, and both oxides recombine to form the silicate [19]. But if the cooling rate is fast enough (quenched) the recombination is avoided, and the product of the dissociation is known as plasma-dissociated zircon (PDZ). During the heating process, the zirconium silicate crystal structure re-arranges into amorphous silica and zirconium oxide phases. On fast cooling, amorphous glassy silica matrix entraps the fine zirconium oxide particles which can be easily disintegrated by acid attack. At high temperature, the stable zirconia phase is the tetragonal zirconia (t-ZrO_2) as inferred from phase diagram of zirconia (**Figure 5**). The resulting tetragonal zirconia during the cooling might transform into monoclinic zirconia which is the stable phase at room temperature [23]. Plasma decomposition of zircon is greatly affected by forward power, plasma and carrier gas flow, and location of power feed port.

The microstructure of PDZ is influenced by the cooling rate after melting which depends strongly on the initial zircon particle size [24]. Thus, relatively slow cooling results in spherulitic crystals of monoclinic ZrO_2 in SiO_2 glass, whereas the rapid cooling gives extremely fine (< 10 nm) crystals of tetragonal ZrO_2 crystals in glassy silica matrix. Major advantages of thermal plasma processing over conventional methods are the ability of plasma reactors to attain high energy density and high temperatures, ability to control the processing atmosphere, increased reaction kinetics, eco-friendly nature of the process, the rapid cooling rate which prevents back reaction, and adaptability to process a variety of materials [25].

Plasma thermal dissociation processes have undergone a variety of process changes to obtain zirconia with high purity and definite crystal structure. Low-power transferred arc plasma (TAP) in which air is used as plasma gas instead of conventional argon gas renders the process cost-effective [24]. Carbothermal plasma dissociation in which coke is added to zircon during plasma dissociation is another adaptation. Energy consumption and overall operating cost of production are drastically reduced by these processes compared to conventional methods which are energy-intensive and expensive [21]. In flight, removal of silica is achieved in carbothermal plasma dissociation [26]. The thermally aided dissociation process becomes a thermo-chemically driven reaction at a much lower temperature. Another major feature is that the oxides of Al, Fe, and Ti and Mg that are present as impurities in zircon mineral are completely removed as vapor during plasma processing.

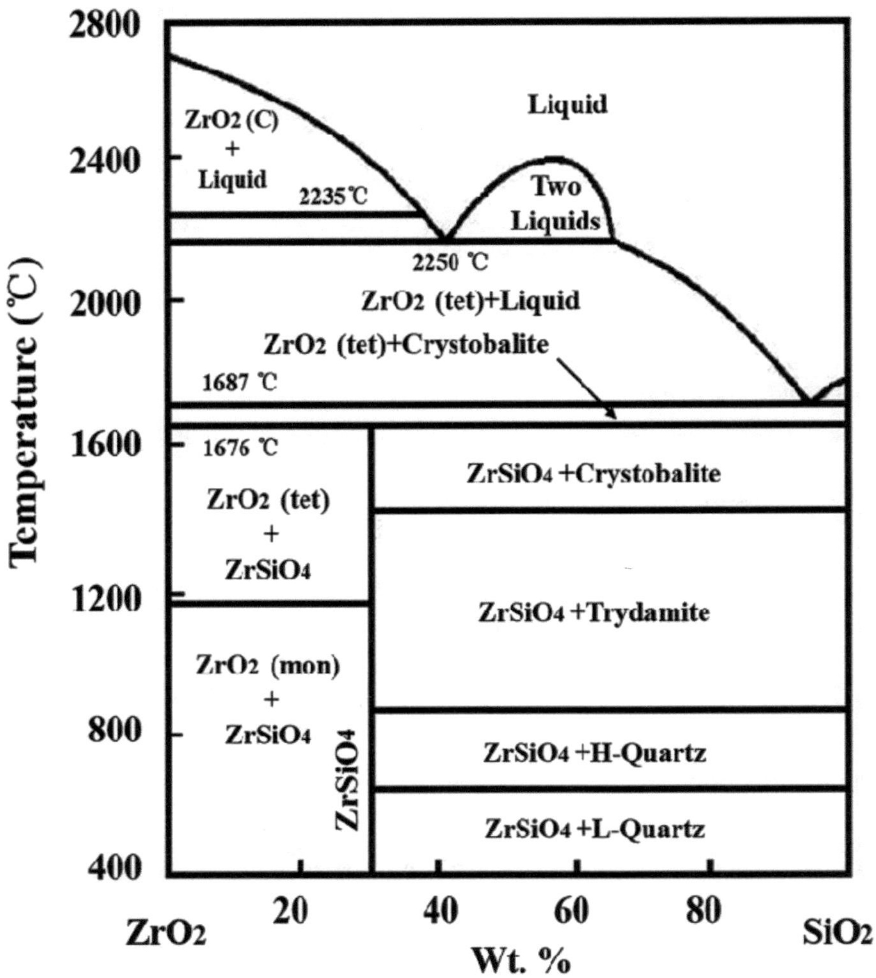

Figure 5.
Phase diagram of zircon [23].

3. Alternate methods for the synthesis of zirconia

The above-mentioned procedures are used generally for the synthesis of macro- or micro-crystalline zirconia which is mainly used as ceramics due to its hardness, abrasiveness, high melting point, and low frictional resistance. At nanoscale, it becomes immensely valuable owing to its high thermal stability, luminescence, refractive index, chemical stability, high specific area, biocompatibility, and ability to exhibit significant antibacterial, antioxidant, and antifungal properties. Such outstanding characteristics have motivated the scientific community to explore zirconia-based nanomaterials in a wide range of technological fields as functional materials, *viz.*, catalysts [27], sensors [28], semiconductor devices [29], ceramics [30], and implants [31]. Apart from that, it can also be employed as a dielectric, electro-optic, and piezoelectric material due to its favorable optical and electrical properties. Zirconia nanoparticles are synthesized by chemical and physical methods.

DOI: http://dx.doi.org/10.5772/intechopen.111737

3.1 Chemical methods for the synthesis of nano-zirconia

Important chemical methods used for the synthesis are solution combustion synthesis (SCS), sol–gel synthesis, hydrothermal synthesis, and co-precipitation.

3.1.1 Solution combustion synthesis (CSC)

Solution combustion synthesis (SCS) has been widely used for the synthesis of metal oxides with desired morphology. This involves the complex exothermic reactions in solution between fuel which is an organic compound and zirconium metal solution. SCS is a self-sustained redox reaction initiated by a source of energy (thermal or electric) between a fuel and an oxidant (usually metal nitrates) [32]. The fuel is typically composed by organics containing carbon and hydrogen that facilitate the liberation of heat by the formation of CO_2 and H_2O during the combustion process. The metal precursors are the source of the metal cations which decides the final metal oxide. One of the methods for the synthesis of zirconia nanopowders is by annealing stochiometric mixtures of zirconium oxy nitrate hydrate in alcohol and urea [33]. Nanocrystalline zirconium oxide powder has been prepared by the glowing combustion method using sucrose as the fuel and zirconyl nitrate as an oxidant in aqueous solution [34]. Solution combustion synthesis is an attractive technique for the preparation of ZrO_2 nanopowders and thin films, owing to its simplicity, energy and time savings, cost-effectiveness, versatility, and higher purity compared to conventional methods [35].

3.1.2 Sol–gel synthesis

Sol–gel synthesis is the stepwise formation of metal oxide nanoparticles in which the sol or colloidal solution of solid particles formed first is transformed into a gel which is an interconnected network of polymerized metal oxide in a solvent. Drying of the gel results in aerogels having nanocrystalline size [36]. For the preparation of zirconia by sol–gel process, zirconium alkoxide in alcohol (precusor) is hydrolyzed with ammonia. Resultant sol is stirred continuously until the gel is formed. The gel formed is dried in conventional oven or by supercritical drying methods. Supercritical drying produces aerogels with smaller size compared to conventional drying methods:

$$\mathrm{Zr\ (OC_3H_7)_4 + C_3H_7OH + NH_4OH \rightarrow Zr\ (OH)_4\ (sol) + C_3H_7OH + NH_3.} \tag{10}$$

$$\mathrm{Zr\ (OH)_{4(Sol)}\ (on\ polycondensation) \rightarrow \left[Zr(OH)_4\right]_n.xC_3H_7OH.xH_2O.} \tag{11}$$

$$\mathrm{\left[Zr(OH)_4\right]_n.xC_3H_7OH.xH_2O\ (110°C) \rightarrow Zr(OH)_4 + C_3H_7OH + H_2O.} \tag{12}$$

$$\mathrm{Zr\ (OH)_4\ (400°C) \rightarrow ZrO_{2(T)}\ (500 - 700°C) \rightarrow ZrO_{2(T)} + ZrO_{2(M)}} \tag{13}$$

In this method, the reactivity of the precursors can be modified using chelating agents, which influence the gelation and ultimately the modification of the size and shape of the particles. The different chelating reagents that have been employed for the preparation of nano-zirconia are acetylacetone, acetic acid, citric acid, and different sugars (e.g., sucrose, maltose, and glucose). The advantages of sol–gel synthesis over other techniques are the ease in controlling the purity, homogeneity, and physical characteristics at low temperature [37].

3.1.3 Hydrothermal synthesis

Hydrothermal method for the synthesis of nanocrystalline zirconia is usually carried out by heating a mixture of precursors (zirconium compounds) with appropriate reagent which are reducing/oxidizing/hydrolyzing in a suitable solvent in a closed vessel [31, 38]. This method ensures the production of nanocrystals with definite morphology, crystal structure, stability, and size. In some methods, surfactants are added to the mixture to avoid agglomeration. Agglomeration can be further eliminated by hydrothermal corrosion methods. This method involves the use of corroding mediums such as sulfuric and hydrochloric acid to break down the hard agglomerate to dispersed fine nanoparticles. By carefully adjusting the reaction parameters, zirconia with definite crystalline structure and required size can be obtained. By introducing suitable dopants at the synthesis stage, zirconia with different physical properties can be synthesized. By introducing Eu^{3+} during the synthesis of cubic zirconia, white light-emitting nano-zirconia is produced which can be used in light-emitting diodes and electronic flashes [39]. Zirconia nanocrystals with different size can be synthesized by careful control of reaction parameters and concentration of reactants. Spindle–/rod-like structures are synthesized by this method [40]. Surfactant-assisted hydrothermal synthesis of zirconia produces thermally stable zirconia crystals with different morphology [41]. Surfactants such as cetyltrimethylammonium bromide (CTAB), sodium dodecyl sulfate (SDS), and Triton X-100 are generally used for this type of synthesis. By tuning the reaction conditions during synthesis, definite phase composition and morphology can be achieved (**Figure 6**).

3.1.4 Microwave-assisted hydrothermal synthesis

This is another method for the synthesis of nano-zirconia. The advantages of this method over the other methods are volumetric heating (the entire volume of solution is evenly heated, instead of relying on heat diffusion processes across the reaction vessels) and short reaction time (reaction time can be as short as a few minutes) [42]. Moreover, the accurate control of morphology and crystal structure is possible by adjusting microwave parameters. Different zirconia phases can be obtained by adjusting pH, temperature, reaction time, pressure, and the precursor used [43]. In this method, microwave digestion of stochiometric composition of zirconia percussor,

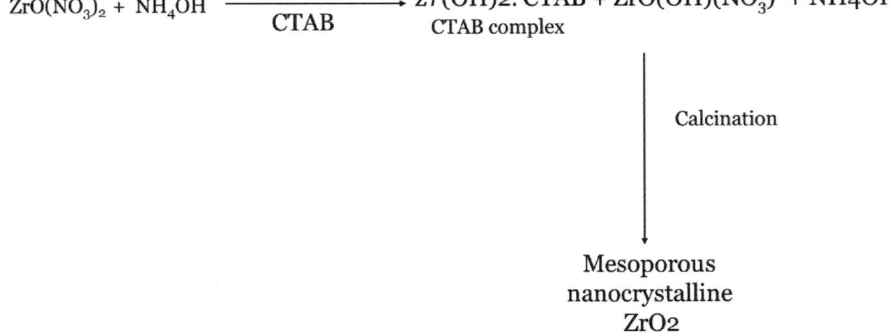

Figure 6.
Surfactant-assisted nano-zirconia synthesis [41].

generally a zirconyl compound (e.g., zirconyl chloride, zirconyl hydroxide, zirconyl nitrate hydrate, and zirconium alkoxides) and an alkali generally sodium hydroxide is carried out at specified temperature and pressure.

3.1.5 Co-precipitation technique

This is an easy method for the synthesis of zirconia nanoparticles. Co-precipitation method is a promising alternative to other methods due to its inherent simplicity, ecological compatibility, precise stoichiometry, structural control, and large-scale production. However, size, shape, and dispersion of powders prepared by this method depends strongly on precipitants used [44]. Hence, the selection of appropriate precipitating reagent is the most important factor in co-precipitation method. Ammonium hydroxide (NH_4OH), ammonium bicarbonate (NH_4HCO_3), ammonium carbonate [$(NH_4)_2CO_3$], sodium hydroxide (NaOH), and urea are the typical precipitants used in precipitation method. pH of the precipitating medium has significant influence on the homogeneity and composition of the precipitate [45]. Precipitants are added to zirconium precursor, typically zirconium oxychloride or nitrate at a controlled rate. As the critical solution concentration of the zirconium hydroxide is reached, nucleation starts followed by growth phase. Zirconium hydroxide on calcination gives zirconia nanoparticles. Morphology of the nanoparticle obtained can be controlled by the ratio of the reactants and calcination temperature [46]. Surface area of the zirconia prepared by this method depends on nature of the precursor, rate of addition of base, pH of the solution, and presence of surfactants and organic anions like oxalate and citrate [47]. At an increased calcination temperature, size of the crystal and its agglomeration increased.

3.2 Solid-phase sintering

In this method, precursors are mixed and milled in a high-intensity ball mill. Mesoporous zirconia with worm-like structure is obtained by milling zirconyl chloride mixed with block polymer surfactant along with sodium hydroxide. The mixture is autoclaved and allowed to crystallize [48]. Sintering method is used for the densification of the material. Preparation parameters have strong influence on the structure of zirconia synthesized. Pore structure changes from microporous to mesoporous with change in parameter. A controlled grain size with good densification can be achieved by proper selection of heating schedule. Densification of powders prepared by chemical vapor methods can undergo one step sintering at 1000°C. Powders prepared by solution technique cannot be sintered at a lower temperature. Sintering process depends on mass transport mechanism [49]. Four underlying mechanisms of sintering are surface diffusion, spread throughout volume boundary, evaporation, and condensation. Temperature and residence time of the material under sintering are decisive in particle size and bond strength. The smaller the particle size, greater will be the bond strength due to the increased contact between the particles (**Figure** 7).

3.3 Biosynthesis of zirconia

Biosynthesis is a greener approach to zirconia synthesis which helps to enhance its biocompatibility and reduce the environmental concerns. This process involves the synthesis of zirconia nanoparticles by reducing, stabilizing, and capping of the metal precursor using natural and renewable agents like microbes and plant parts [51–55].

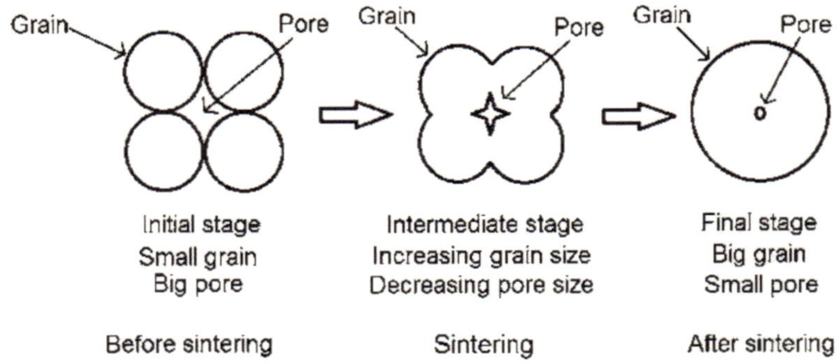

Figure 7.
A model showing sintering process [50].

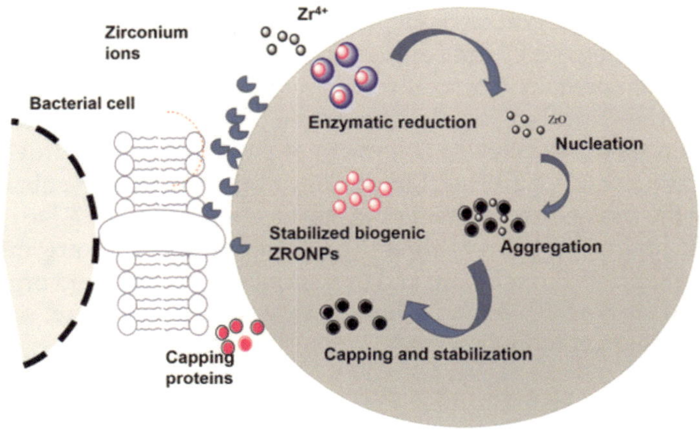

Figure 8.
Biosynthesis of zirconia [56].

Major steps involved in this process are reduction and stabilization. There are intracellular and extracellular pathways adopted for reduction. In intracellular pathway, metal ion is transported into the cell by interaction with the negative receptors of the cell wall. The enzymes of the cell cause the reduction of metal. In extracellular synthesis, zirconium ions are reduced by an enzyme – nitrate reductase. After the reduction, the zirconia nanoparticles are formed *via* nucleation, aggregation, and growth (**Figure 8**). The protein species released by the cell wall act as capping agents which stabilize zirconia nanoparticles [56].

4. Characterization of zirconia

Characterization is the essential requirement to understand the physical, chemical, structural, and morphological properties of a newly synthesized material. Characterization involves the identification of different components present in the material, their quantification, structural elucidation, and identification of morphological properties of the material. Quantification of different components in zirconia is carried out by chemical procedures and by instrumental techniques like X-ray photoelectron spectroscopy (XPS), wavelength-dispersive X-ray fluorescence spectroscopy

(WD-XRF), and laser ablation inductively coupled plasma mass spectrometry (LA-ICP-MS). Structural elucidation is carried out by instrumental techniques like X-ray diffraction (XRD) spectroscopy, Fourier-transformed infrared spectroscopy (FTIR), N2 adsorption measurements, and Raman spectrometry. Morphology of synthesized material is usually assessed with scanning electron microscopy (SEM), transmission electron microscopy (TEM), and atomic force microscopy (AFM).

4.1 Chemical characterization of zirconia

Complete chemical characterization of zirconia is the quantification of elements present in zirconia. Multielement analysis of zircon samples is usually carried out by inductively coupled plasma atomic emission spectroscopy (ICP-AES) or inductively coupled plasma mass spectroscopy (ICP-MS) after the quantitative dissolution [57]. Zirconia being a highly refractory material, quantitative decomposition of the material cannot be accomplished by usual acid decomposition procedures. Fusion of zirconia with sodium carbonate, sodium metaborate, and lithium metaborate are reported methods for complete dissolution. Microwave-assisted digestion of zirconia with a mixture of hydrofluoric acid, sulfuric acid, and hydrochloric acid is also used for decomposition of zirconia. Once in solution state, the concentration of elements is determined by either ICP-AES or ICP-MS. Electrothermal vaporization inductively coupled plasma mass spectrometry (ET-ICP-AES) in presence of a modifier is an effective method for the characterization of high-purity zirconia [58]. ICP-MS has also been used for the single-particle analysis of zirconia colloids in water [57]. Laser ablation inductively coupled plasma mass spectrometry is an excellent technique for the chemical characterization of zirconia samples.

4.2 X-ray photoelectron spectroscopy

Surface chemical characterization of zirconia is carried out using XPS. XPS also known as electron spectroscopy for chemical analysis (ESCA) is a quantitative spectroscopic technique which gives valuable information on the surface elemental composition and chemical state of the components of the material under study [59]. XPS provides surface analysis data and the average depth of analysis for an XPS measurement is approximately 5 nm. XPS is typically accomplished by exciting a samples surface with mono-energetic Al kα or Mg Kα X-rays which prompts the emission of photoelectrons from sample surface. An electron energy analyzer is used to measure the energy of the emitted photoelectrons. From the binding energy and intensity of a photoelectron peak, the elemental identity and chemical state of the element can be determined. Surface defects can be inferred from chemical state of the elements on the surface. Wide-scan spectrum, core-level spectrum, and the upper valence band (UVB) spectrum are the major analysis modes used for characterization in XPS. The wide-scan spectrum gives elemental information of the surface. **Figure 9a** shows the wide-scan spectrum of zirconia. It shows peaks corresponding to Zr and oxygen. Presence of any peak other than that of Zr and O_2 shows contamination or doping. **Figure 9b** shows the core-level spectrum of zirconia. XPS core-level peaks are used to obtain information on the chemical state of the surface, and crystalline phase. **Figure 10** gives the Zr-$3d_{5/2}$, Zr-$3d_{3/2}$, and O-1 s core-level peaks of thin films of zirconia with different oxygen content. The chemical shift observed in the films with higher oxygen content corresponds to different oxidation state of zirconia. Peak shift in core-level peaks, Zr-$3d_{5/2}$, Zr-$3d_{3/2}$, and O-1 s of **Figure 11** corresponds to phase change.

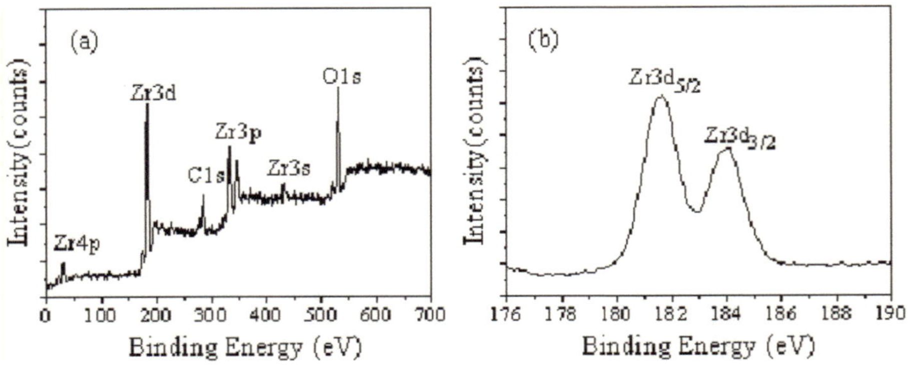

Figure 9.
The wide-survey scan spectrum (a) and Zr 3d core-level XPS spectra (b) of zirconia [60].

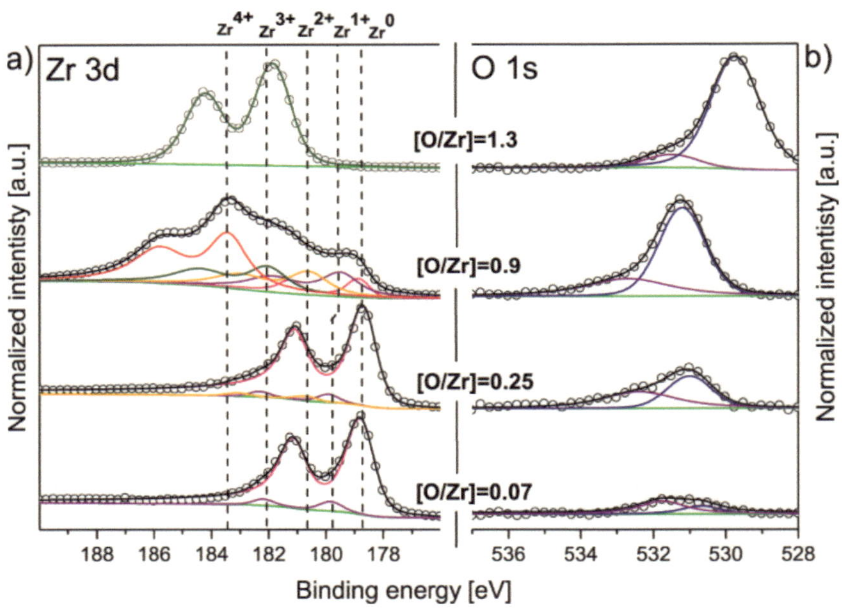

Figure 10.
(a) Core-level Zr 3d (XPS) spectra of bulky metallic ([O/Zr] = 0.07), partially oxidized films ([O/Zr] = 0.25 and [O/Zr] = 0.9) thin oxide films([O/Zr] = 1.3) with different oxidation states: Zr^0 for metallic, Zr^{1+}, Zr^{2+}, Zr^{3+} for sub-oxides and Zr^{4+} for stoichiometric oxide. (b) Core-level O1s spectra (XPS) for the corresponding Zr3d spectra [61].

Identification of different crystalline phases by shift in core-level peaks is extremely difficult as the variation in crystal structure does not produce large shift in the core levels. However, by analyzing the valence band of XPS, phase transformation data can be obtained. **Figure 12** shows the upper valence band XPS (UVBXPS) spectrum of zirconia surfaces grown by dry thermal oxidation in the temperature range 300–450°K. Oxide films grown below 400°K are predominantly amorphous, and formation of tetragonal phase starts above this temperature. There is a pronounced change in the shape of upper valence band spectra with increase in oxidation temperature. This can be attributed to the gradual change in the formation of tetragonal

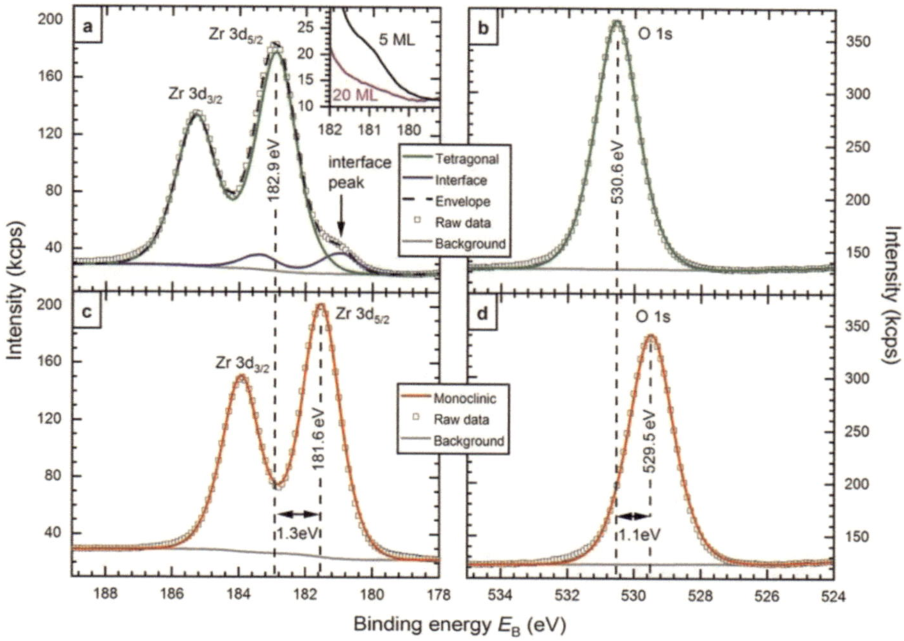

Figure 11.
XPS spectra of (a and b) tetragonal and (c and d) monoclinic zirconia films annealed at 650°c and 850°c [62].

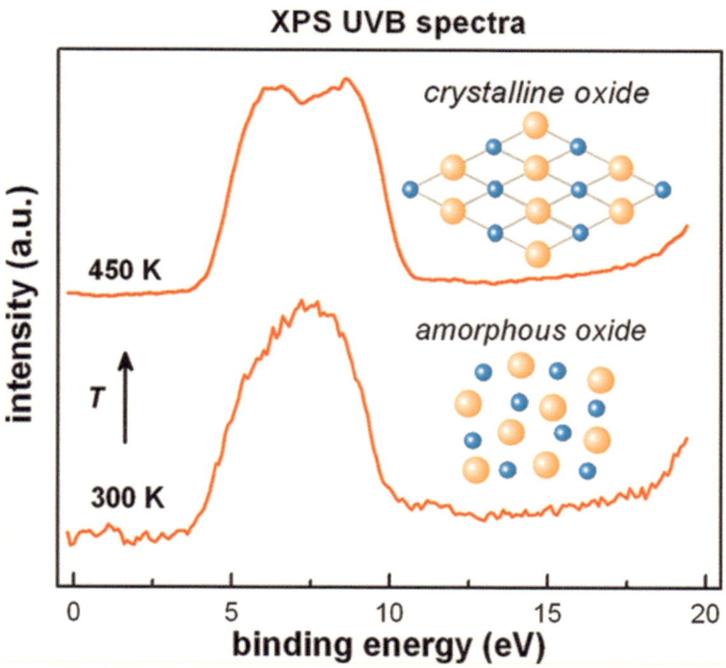

Figure 12.
XPSUVB spectra of thermally grown ZrO_2 films at temperatures 300°K and 450°K [63].

phase (increase in crystallinity) with temperature above 400°K. Change in the shape of the spectra with crystallinity is due to the increase in Zr-O bond ionicity and changes in the first coordination spheres of both Zr and O.

4.3 Fourier-transformed infrared spectroscopy (FTIR)

FTIR spectrum is an effective tool for the identification of functional groups present in a molecule [64]. When infrared radiation passes through a molecule, it causes changes in the dipole moment of the molecule which corresponds to a definite vibrational energy. Since every functional group is composed of different atoms and bonds with different bond strengths, frequencies of vibrations are unique to individual groups and classes of functional groups (e.g., O-H and C-H stretching frequency appear around 3200 cm^{-1} and 2900 cm^{-1}, respectively). Since the collection of vibrational energy bands for all the functional groups of a molecule is unique to every molecule, these peaks can be used for identification using library searches of comprehensive sample database. FTIR spectrum of ZrO_2 nanoparticles and surfactant-modified zirconia are shown in **Figure 13**. The absorption peak at 608 cm^{-1} corresponds to Zr-O stretching vibrations. A shift in the position of this peak is observed in the surfactant-modified zirconia. Two extra broadbands at 1607 cm^{-1} 1nd 1400 cm^{-1} in the spectra of surfactant modified ZrO_2 are attributed to the bending vibration of C-H bonds.

4.4 BET surface area analysis

Brunauer–Emmett–Teller (BET) surface area analysis is used to measure total surface area, total pore volume, and pore size distribution of a material [64]. In this method, multipoint analysis of analytes-specific surface area (m^2/g) is measured through gas absorption analysis. When inert gases like N2 is flown over the solid sample, the gas molecules absorb on to analyte surface and in pores due to the weak van der Waals forces forming a monolayer of gas on the surface. After the adsorption layers are formed, the sample is removed from the nitrogen atmosphere and heated to release the adsorbed nitrogen from the material and quantified. The data obtained is used to

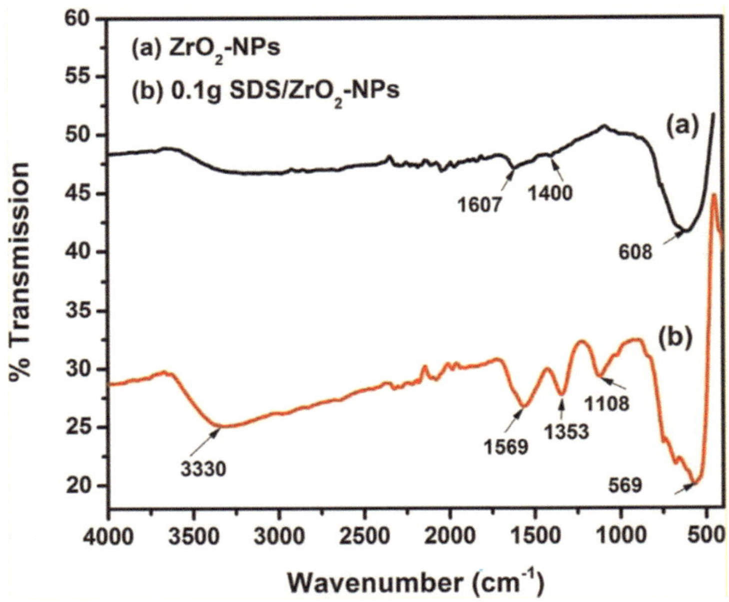

Figure 13.
FTIR spectra of zirconia nanoparticles (a) and surfactant-modified zirconia nanoparticles (b) from 4000 to 400 cm^{-1} [65].

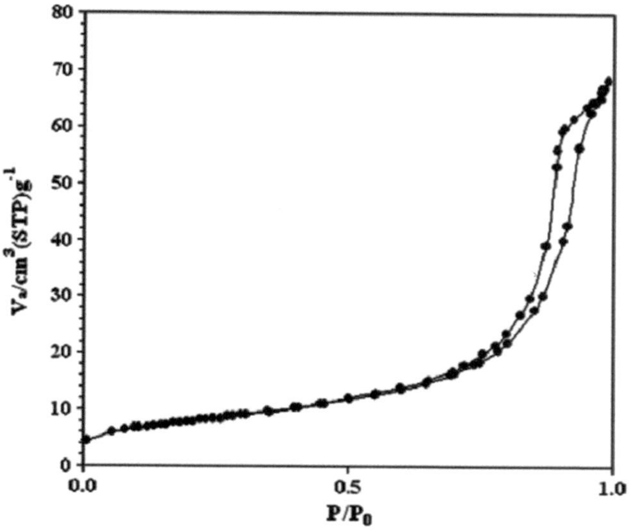

Figure 14.
BET adsorption desorption curve for zirconia nanoparticles [66].

plot the BET isotherm which is a plot of amount of gas adsorbed as a function of pressure. At normal temperatures, the interaction between solid and gases is minimal, and the sample surface is cooled with liquid nitrogen to obtain appreciable adsorption. Surface area is calculated using BET equation from the adsorption branch of isotherm, and pore size distribution was calculated from desorption branches using the Barrett–Joyner–Halenda (BJH) method [64]. **Figure 14** shows the BET adsorption–desorption curve of ZrO_2 nanoparticles. The surface area and size of the nanoparticle obtained from this curve are $44m^2/g$ and 24 nm, respectively [66].

4.5 X-ray diffraction spectroscopy (XRD)

Powder X-ray diffraction is a method used for the identification of crystalline structure and phases of a material [67]. When X-rays interact with crystalline samples, they are diffracted at a particular angle which satisfies Bragg's equation ($n\lambda = 2d\sin\theta$) which gives the relation between wavelength of electromagnetic radiation to the diffraction angle and the lattice spacing in a crystalline sample. The diffracted X-rays are detected, processed, and counted. By scanning the sample through a range of 2θ angles, all possible diffraction directions of the lattice are obtained due to the random orientation of the powdered material. Conversion of these diffraction peaks to d spacing by comparing it with standard reference patterns allows the identification of crystalline structure. Quantitative characterization of lattice defects and relative stress endured by the crystals can also be elucidated by XRD. As zirconia exists in three different crystalline structures, ascertaining the crystalline phase of the newly synthesized material is the most important part of the characterization of zirconia. Crystalline structure of synthesized zirconia depends on so many factors like temperature of calcination, pH of synthesis, and the reagents used. **Figure 15** shows ZrO_2 nanoparticles synthesized at different temperatures (400, 500, and 600°C). The well-defined peaks at 2θ = 30.2, 35.0, 50.4, 60.0, and 62.7°correspond to the diffractions of the (101), (110), (200), (211), and (202) crystalline planes of cubic zirconia. The shape of the peaks at 35.0, 50.4, and 60.0° are slightly asymmetric, suggesting the

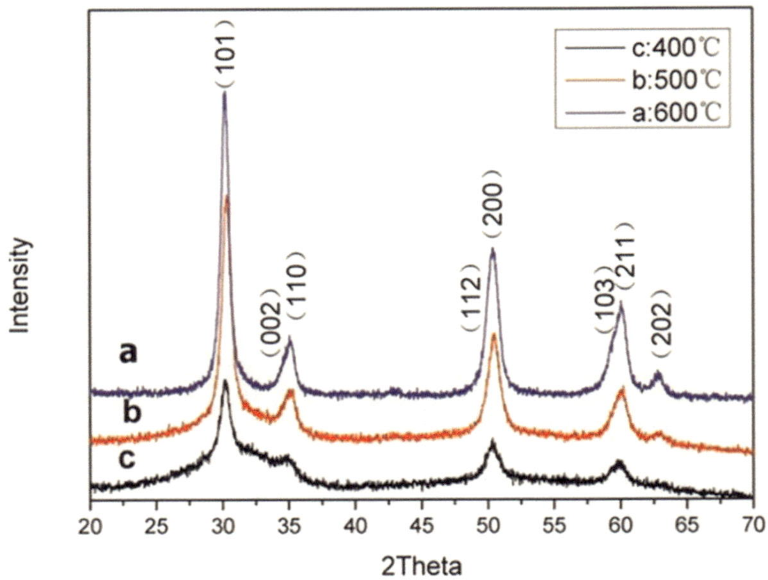

Figure 15.
XRD patterns of ZrO_2 NPs synthesized at different temperatures [68].

formation of a tetragonal phase. The shoulder at 2θ = 34.5° is due to the diffraction of the (002) crystalline plane of tetragonal phase zirconia. The appearance of peaks corresponding to tetragonal and cubic phase in the XRD spectrum on increasing the temperature from 400 to 600°C gives a clear indication of the conversion of amorphous to crystalline form with increase in temperature.

4.6 Raman spectroscopy

Raman spectroscopy is a powerful, nondestructive, user-friendly technique to determine the surface characteristics and subsurface damages. This technique is based on Raman effect which in-elastic scattering process which occurs when an incident photon interacts with phonons in the material [69]. When the photons of a particular wavelength interact with the vibrating phonons of the materials, the light is scattered at a frequency and the difference in frequency between the incident and the scattered light provides the information of the lattice vibrations. Raman vibrational spectra provides a structural finger print for the molecular identification. Raman spectra of the molecules are widely influenced by the microstructural changes and impurities in the molecule. Hence, the information from Raman spectra like band position, shift, and intensity can be used to characterize the defects and subsurface damages induced by crystal growth. Raman spectra obtained from the instrument gives information on molecular vibrations and crystal structure of materials [70]. Raman spectra of the crystalline molecules satisfy energy conservation rule, wave vector conservation, and polarization selection rule. But if there are some defects and damages in the crystal, this will induce a reduction in symmetry and breakdown of the selection rules will take place. This may give rise to fresh bands in the Raman spectra or broadening of spectra. A change in the Raman spectra is observed when a material is strained or crystal structure of the material is changed [71].

Raman peaks are widely used to identify different crystalline phases of zirconia.

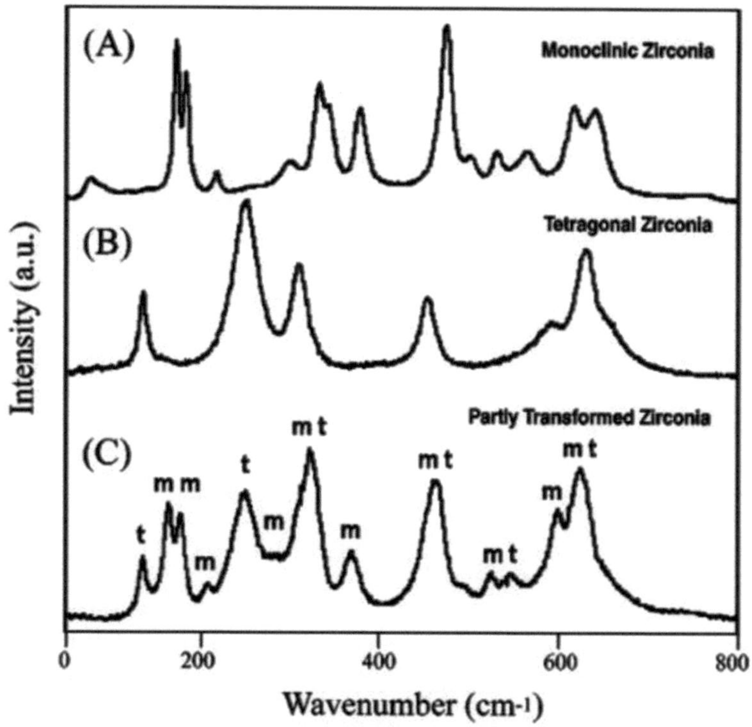

Figure 16.
Raman spectra of different phases of zirconia [72].

Figure 16 shows the Raman spectra of monoclinic (A), tetragonal (B), and partly tetragonal partly monoclinic (C) zirconia phases. The two sharp peaks at 142 cm^{-1} and 256 cm-[1] corresponds to tetragonal phase. Another two bands at around 316 cm^{-1} and 460 cm^{-1} also correspond to tetragonal phase. The monoclinic phase shows two sharp peaks at 178 cm^{-1} and 190 cm^{-1}. Broadband at 384 cm-[1] also corresponds to monoclinic structure. Partially transformed zirconia of an overlap of bands corresponding to tetragonal and monoclinic phases are observed. Monoclinic phase content (*Vm*) of zirconia samples can be evaluated by the following formula which was first proposed by Clarke and Adar [73]:

$$\mathrm{Vm} = \mathrm{Im178} + \frac{\mathrm{Im189}}{0.97(\mathrm{It145} + \mathrm{It260})} + \mathrm{Im178} + \mathrm{Im189},$$

where *Im* and *It* are the intensity of the peaks corresponding to monoclinic and tetragonal phases at different wavenumbers, respectively.

4.7 Scanning electron microscopy (SEM)

Scanning electron microscope is a powerful tool for the two-dimensional topographical imaging. The high-energy electrons emitted from the surface of a material after being exposed to a highly focused beam of electrons from an electron gun are used to produce high-definition images with a resolution of 20 to 0.4 nm [74]. There are two modes electron analysis each one giving different information. Backscattered electrons give contrast based on the different chemical composition across an image. Secondary electrons which are emitted close to the surface of the sample give information on surface topography. Most of the SEM instruments are equipped with energy-dispersive X-ray spectroscopy which is based on the characteristic of X-rays emitted (which is unique to each element) by the element when exposed to an electron beam. A semi-quantitative data of chemical composition of the sample is obtained by this method. Surface morphology of zirconia synthesized by different methodologies has different characteristics which are ascertained by SEM data. **Figure 17** shows the SEM data of zirconia with EDS spectra. Peaks in EDS spectra give qualitative data of the material synthesized. Energy-dispersive X-ray spectroscopy (EDS) spectra of pure zirconia show the peaks corresponding to Zr and oxygen. Presence of other peaks can be attributed to dopants, impurities, and incompletely removed reagents. **Figure 18** shows the agglomerated zirconium oxide powder. **Figure 19** shows SEM data of zirconia before and after ball milling at two different magnifications. Surface characteristics of the two are drastically different at 50-μm magnification. However, magnification at 300 nm shows

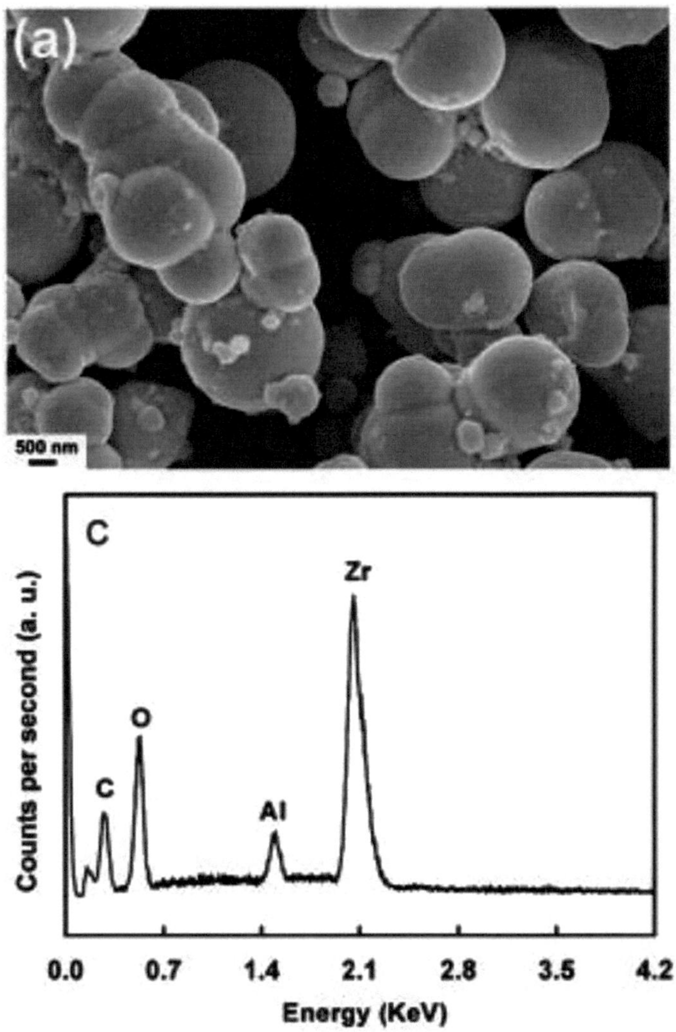

Figure 17.
SEM image and EDX spectra of freshly synthesized zirconia [75].

Figure 18.
SEM picture of agglomerated zirconium oxide powder [76].

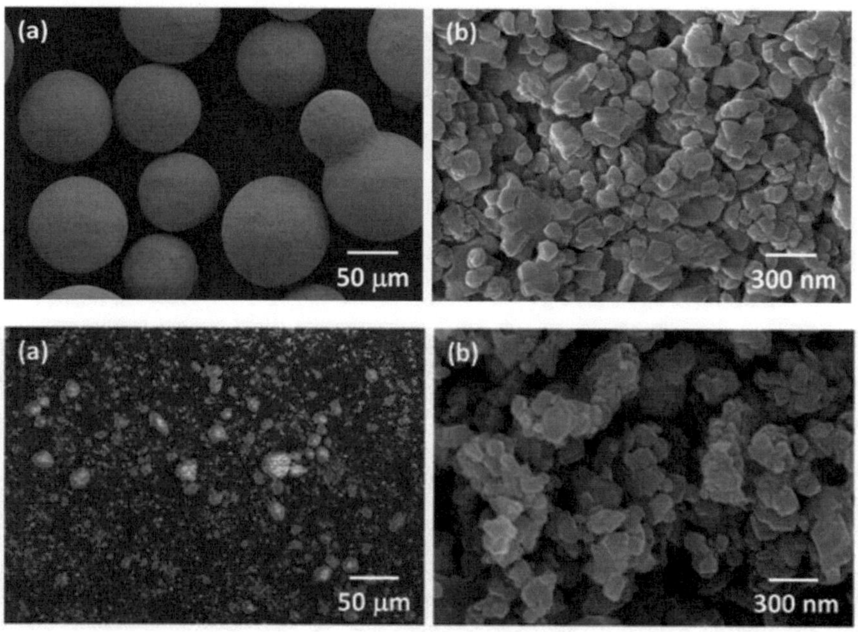

Figure 19.
SEM picture of zirconia before (a) and after (b) ball milling [77].

essentially same surface characteristics. It is inferred from this data that morphology of zirconia remains unaffected even after ball milling.

4.8 Transmission electron microscopy (TEM)

Transmission electron microscopy data provides topographical and morphological information about specimens using energetic beam of electrons [78]. When a high-energy electron beam is allowed to pass through a thin slice of a material, the electrons interact with atoms of the material. TEM offers a powerful magnification of the order of a million times. A resolution up to 0.05 nm can be achieved by TEM. The highly detailed images provide valuable insight into elemental and compound structure, leading to provide information on surface features, shape, size, and structure. TEM offers valuable information on the inner structure of the sample. **Figure 20** shows the TEM data of nanocrystalline zirconia synthesized by sol–gel method. **Figure 21** is the in situ TEM data of zirconia under heavy ion irradiation. Crystalline materials can be

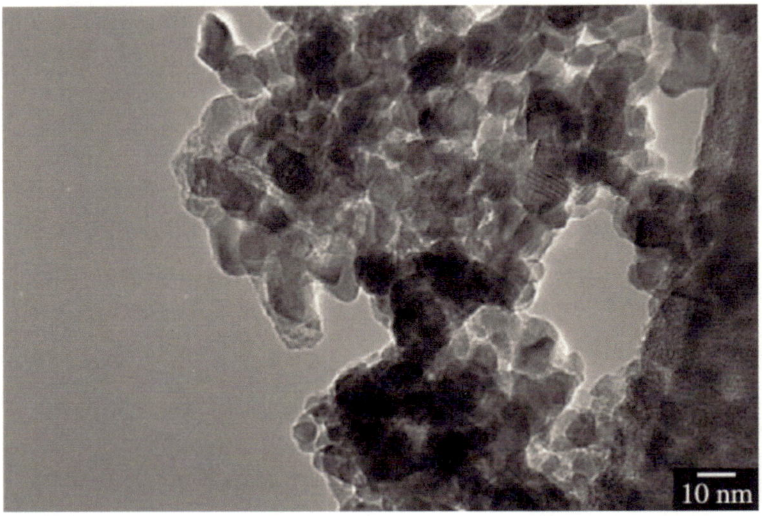

Figure 20.
TEM image of zirconia nanoparticle synthesized by sol–gel method [79].

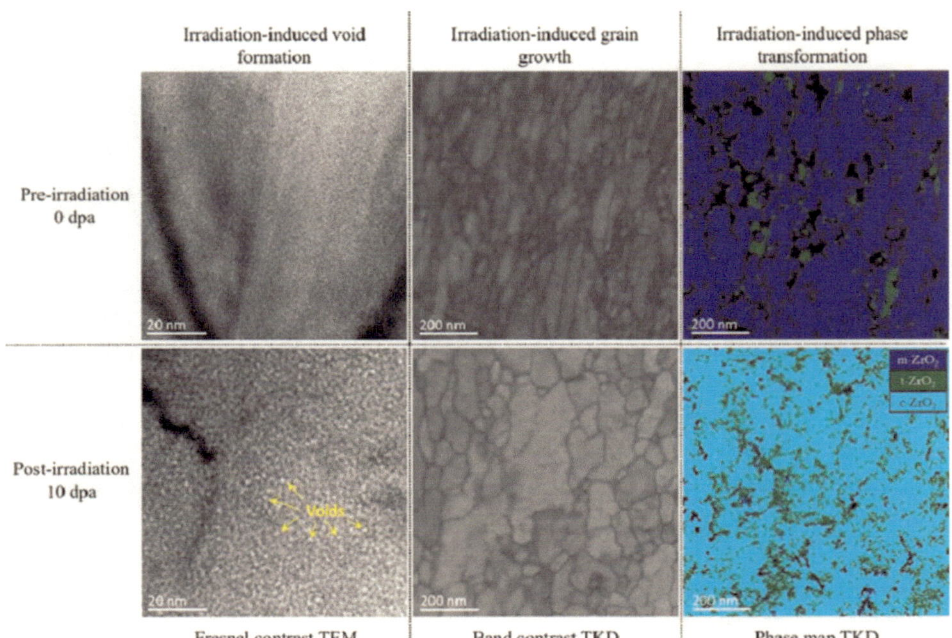

Figure 21.
In situ TEM images of ZrO_2 before and after ion irradiation [80].

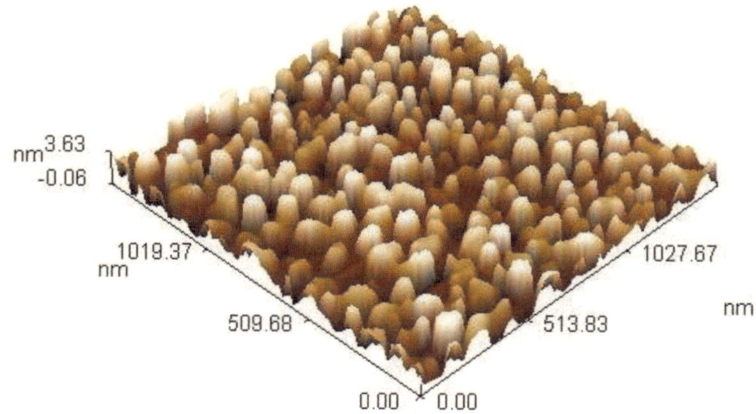

Figure 22.
Particle size and distribution zirconia nanoparticles by AFM [81].

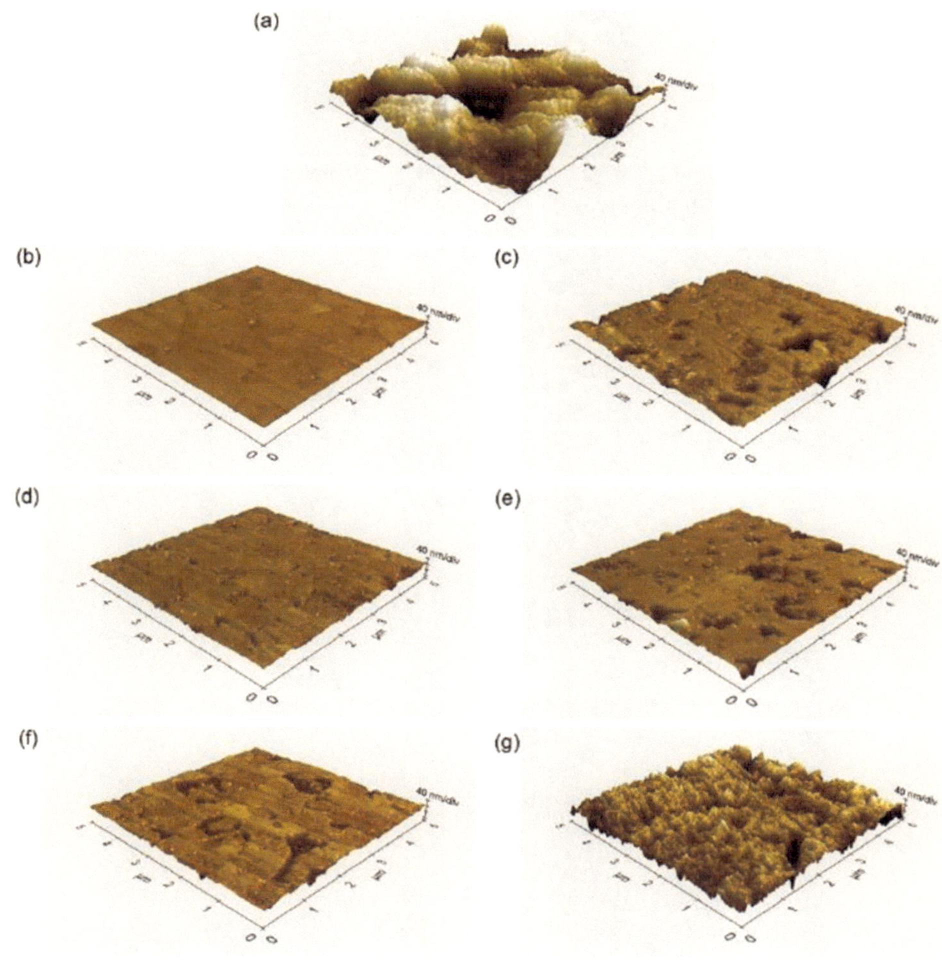

Figure 23.
AFM images of zirconia ceramics after different surface treatments. (a) APA, (b) 10F5, (c) 10F30, (d) 20F5, (e) 20F30, (f) 30F5, and (g) 30F30.

made amorphous by ion irradiation. Appearance of voids and phase transformation is observed in the TEM data after ion irradiation.

4.9 Atomic force microscopy (AFM)

Atomic force microscopy is a powerful tool to characterize the surface of a material down to atomic scale. AFM can be used to obtain nanoscale chemical, mechanical, electrical, and magnetic properties. AFM offers the three-dimensional visualization of individual particles and group of particles. This is a very good tool to elucidate the surface roughness and grain size and shape of zirconia nanoparticles and ceramics. **Figure 22** is an AFM image that shows shows the size of three-dimensional arrangement of zirconia nanoparticles [81].

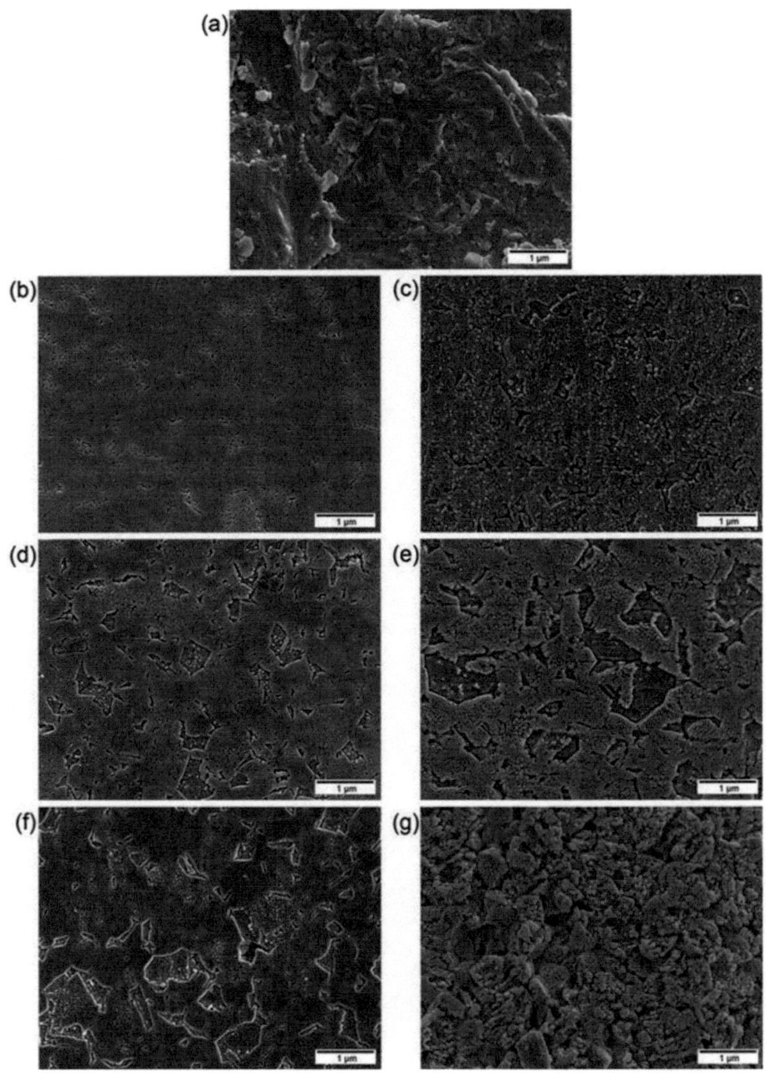

Figure 24.
SEM images of zirconia ceramics after different surface treatments. (a) APA, (b) 10F5, (c) 10F30, (d) 20F5, (e) 20F30, (f) 30F5, and (g) 30F30.

Tetragonal zirconia used in dental implants are coated with resin. Adhesiveness of the resin to zirconia increases with roughness of ceramic surface. Surface roughness of zirconia ceramics can be increased by either physical abrasion or chemical etching with hydrogen fluoride (HF). AFM pictures of zirconium ceramic material after different surface treatments are given in **Figure 23** [82]. AFM pictures shows changes in roughness with different treatments like (a) abrasion with alumina, (b) etching with 10%HF for 5 minutes, (c) etching with 10%HF for 30 minutes, (d) etching with 20%HF for 5 minutes, (e) etching with 20%HF for 30 minutes, (f) etching with 30% HF for 5 minutes, and (g) etching with 30%HF for 30 minutes. Etching with 30%HF for 30 minutes gave the roughness required for proper adhesiveness of resin to ceramic. SEM images complement AFM data (**Figure 24**). Data obtained by AFM images are supported by the SEM data. SEM images shows the formation of micro and nanopores on the surface by different treatments.

5. Conclusion

The main source of zirconia, the widely used material in the field of ceramics, electronics, and refractories is zircon, an accessory mineral of placer deposits. Alkali fusion and plasma thermal dissociation are the important decomposition methods adopted widely to convert zircon to zirconia. Various methodologies are adopted for the synthesis of zirconia from its zirconium precursors depending on the particle size and crystal structure. Solution combustion synthesis, sol–gel synthesis, hydrothermal synthesis, and solid-phase sintering are some of the methods adopted for synthesis of different types of zirconia material. Depending on the utility, the synthesized zirconia further undergoes surface modification techniques like surface cleaning and roughening. Doping of zirconia with other materials for increasing the mechanical strength, electrical conductivity, and biocompatibility have been carried out during or after the synthesis. Characterization of synthesized material is essential for ascertaining properties of the synthesized material. Elemental characterization of zirconia can be best accomplished by the destructive solution techniques such as ICP-AES and ICP-MS or by nondestructive solution techniques such as LA-ICP-MS. XPS also can be used for elemental characterization. However, the data obtained pertain only to the surface, but XPS core data can also be used to check the surface chemical state of the zirconium and oxygen. XPS UVB data gives some idea about the crystallinity of the material. Different crystal structure of the synthesized zirconia can be ascertained using XRD data which can be complimented with Raman studies. FTIR data are used to analyze the sample for different functional groups and bonding in the sample. SEM, TEM, and AFM data are used to ascertain the morphology and topography of the prepared material. No single technique is panacea for material characterization, and it is not easy to characterize a material using all the techniques. Hence, judicial choice of combination of techniques for characterization and sufficient knowledge to interpret the data obtained are essential for efficient characterization of a freshly synthesized material.

Author details

Bincy Cyriac
Chemistry Department, Christ Deemed to be University, Bangalore, Karnataka, India

*Address all correspondence to: bincy.cyriac@christuniversity.in

References

[1] Mohammed NA, Daher AM. Preparation of high-purity zirconia from zircon: An anion-exchange purification process. Hydrometallurgy. 2002;**65**(2-3): 103-107. DOI: 10.1016/S0304-386X (02) 00042-7

[2] Han H, Zhu J. Surface science studies on the zirconia-based model catalysts. Topics in Catalysis. 2013;**56**:1525-1541. DOI: 10.1007/s11244-013-0156-5

[3] Blumenthal WB. The Chemical Behavior of Zirconium. Princeton, NJ: D. Van Nostrand Company Inc.; 1958. 58 p.

[4] Selby JH. The industrial uses of zircon and zirconia, and the radiological consequences of these uses. In: Proceeding of International Symposium. Seville, Spain. Vienna: IAEA; 2007. p. 95

[5] Elkady MY, Mohmed AM, Daher AME, Wafaa H, Saleh WH, Negam S, et al. Preparation of La2Zr2O7 ceramic from Egyptian black sand. Journal of Radiation Research and Applied Sciences. 2018;**11**(1):37-42. DOI: 10.1016/j.jrras.2017.10.005

[6] Singh Y. Rare earth mineral deposits in Indian context. Society of Earth Scientists Book Series. Cham, Switzerland: Springer; 2020. pp. 137-178. DOI: 10.1007/978-3-030-41353-8_2

[7] Moussa MO, Kouidri S, Rached H, Haid S, Beloufa N, Ouadha I, et al. Theoretical investigation on the optoelectronic properties of ZrxSi1 xO2 tetragonal hypothetical alloys from zircon family. Applied Physics A. 2022; **128**:231. DOI: 10.1007/s00339-022-05362-x

[8] El Barawy KA, El Tawil SZ, Francis AA. Alkali fusion of zircon sand. Transactions of the institution of mining and metallurgy (section C). Mineral Processing and Extractive Metallurgy. 2000;**109**(1):C49-C56

[9] Yamagata C, Andrade JB, Ussui V, Limai NB NB, José OA, Yamagata PC. High purity zirconia and silica powders via wet process: Alkali fusion of zircon Sand. Materials Science Forum. 2008; **591-593**:771-776

[10] Song J, Fan JF, Liu JC, Liu R, Qu JK, Qi T. A two-step zircon decomposition method to produce zirconium oxychloride: Alkali fusion and water leaching. Rare Metals. 2020;**39**(4): 448-454. DOI: 10.1007/s12598-015-0537-y

[11] Murti CFK, Aldila H, Triwikantoro E. Particle size analysis of the synthesised ZrO2 from natural zircon sand with variation of pH deposition using alkali fusion-coprecipitation method. Material Science Forum. 2019;**966**:89-94. DOI: 10.4028/www.scientific.net/MSF.966.89

[12] Septawendar R, Sutardi S, Karsono U, Sofiyaningsih N. A low-cost, facilem on production of nano zirconia and silica from local zircon in a large scale using a sodium carbonate sintering technology. Journal of The Australian Ceramic Society. 2016;**52**(2): 92-102

[13] El Tawil SZ, El Barawy KA, Francis AA. Cubic zrconia from zircon sand by firing with CaO/MgO mixture. Journal of the Ceramic Society of Japan. 1999;**107**(3):193-198

[14] Zhang J, Wang L, Jiang D. Decomposition process of zircon sand concentrate with CaO-NaOH. Rare Metals. 2012;**31**(4):410. DOI: 10.1007/s12598-012-0529-0

[15] Bidaye AC, Venkatachalam S. Studies on the chlorination of zircon: Part I. static bed investigations. Metallurgical and Materials Transactions. 1999;**30B**:205

[16] Poernomo H, Sajima Pusporini ND. Synthesis of zirconium oxychloride and zirconia low TENORM by zircon sand from landak West Kalimantan. Journal of Physics: Conference Series. 2020; **1436**:012106. DOI: 10.1088/1742-6596/1436/1/012106

[17] Wilks PH, Ravinder P, Grant CL, Pelton PA, Downer RJ, Talbot ML. Plasma process for zirconium dioxide. Chemical Engineering Progress. 1972;**68** (4):82-83

[18] Butterman WC, Foster WR. Zirconium stability and ZrO_2-SiO_2 phase diagram. American Mineralogist. 1967;**52**: 880

[19] Rendtorffabc NM, Suárezabd G, Conconiac MS, Singhe SK, Agliettiabd EF. Plasma dissociated zircon (PDZ) processing; influence of the Zr:Si ratio in the composition, microstructure and thermal recrystallization. Procedia Materials Science. 2012;**1**:337-342

[20] Ananthapadmanabhan PV, Sreekumar KP, Iyer KV, Venkatramani N. Plasma thermal dissociation of Indian zircon. Journal of Alloys and Compounds. 1993;**196**(1–2): 251-254

[21] Yuan Z, Li W, Liu J, Luo M, Liu J. Production of zirconia from zircon using a plasma-rotating furnace. Scandinavian Journal of Metallurgy. 2004;**33**:189-192

[22] Kaiser A, Lobert M, Telle R. Thermal stability of zircon (ZrSiO4). Journal of the European Ceramic Society. 2008;**28**: 2199-2211

[23] Suzuki M, Sodeoka S, Inoue T. Structure control of plasma sprayed zircon coating by substrate preheating and post heat treatment. Materials Transactions. 2005;**46**(3):669-674

[24] Ananthapadmanabhan PV, Lusvarghi L, Yugeswaran S. Zircon dissociation in air plasma through low power transferred arc plasma torch. Ceramics International. 2014;**8**:68. DOI: 10.1016/j.ceramint.2014.08.068

[25] Syamprasad U, Bhattacharjee S, Galgali RK, Mohapatra BK, Mohanty BC. Studies on plasma dissociation of Indian zircon in a specially developed plasma reactor. Journal of Materials Science. 1992;**27**(7):1762-1766

[26] Manikandana S, Kuberana A, Yugeswaranb S, Ananthapadmanabhana PV, Pandiyarajc NK, Shajana XS, et al. Silica-free zirconia from zircon mineral by thermal plasma processing. Materials and Manufacturing Processes. 2021; **36**(2):88-199. DOI: 10.1080/10426914.2020.1832681

[27] Gole JL, Prokes SM, Stout JD, Glembocki OJ, Yang R. Unique properties of selectively formed zirconia nanostructures. Advanced Materials. 2006;**18**:664

[28] Zhuiykov S, Miura N. Development of zirconia-based potentiometric NOx sensors for automotive and energy industries in the early 21st century: What are the prospects for sensors. Sens. Actuators B Chemical. 2007;**121**(2): 639-651. DOI: 10.1016/j.snb.2006.03.044

[29] Wilk RM, Wallace JM, Anthony GD, Wilk GD, Wallace RM, Anthony JM. High-κ gate dielectrics: Current status and materials properties considerations high- gate dielectrics: Current status and materials properties considerations. Journal of Applied Physics. 2001;**89**(10): 5243-5274. DOI: 10.1063/1.1361065

[30] Jin XJ. Martensitic transformation in zirconia containing ceramics and its applications. Current Opinion in Solid State & Materials Science. 2005;**9**: 313-318. DOI: 10.1016/j.cossms.2006.02.012

[31] Gupta S, Noumbissi S, Kunrath MF. Nano modified zirconia dental implants: Advances and the frontiers for rapid osseointegration. Medical Devices and Sensors. 2020;**3**:1-8. DOI: 10.1002/mds3.10076

[32] Carlos E, Martins R, Fortunato E, Branquinho R. Solution combustion synthesis: Towards a sustainable approach for metal oxides. Chemistry - A European Journal. 2020;**26**(42): 9099-9125. DOI: 10.1002/chem.202000678

[33] Branquinho R, Santa A, Carlos E, Salgueiro D, Barquinha P, Martins R, et al. Solution combustion cynthesis: Applications in oxide electronics. In: Developments in Combustion Technology. London, UK: IntechOpen; 2016. DOI: 10.5772/64761

[34] Boopalan K, Vijayaraghavan R, Chidambaram K, Mudali UMK, Raj B. Preparation and characterization of nanocrystalline zirconia powders by the glowing combustion method. Journal of the American Ceramic Society. 2010; **93**(11):3651-3656. DOI: 10.1111/j.1551-2916.2010.03947.x

[35] Matias ML, Carlos E, Branquinho R, do Valle H, Marcelino J, Morais M, et al. A comparison between solution-based synthesis methods of ZrO2 nanomaterials for energy storage applications. Energies. 2022;**15**:6452. DOI: 10.3390/en15176452

[36] Lim HS, Ahmad A, Hamzah H. Synthesis of zirconium oxide nanoparticle by sol-gel technique. AIP Conference Proceedings. 2013;**1571**:812. DOI: 10.1063/1.4858755

[37] Suciua C, Gageab L, Hoffmanna AC, Moceanb M. Sol–gel production of zirconia nanoparticles with a new organic precursor. Chemical Engineering Science. 2006;**61**:7831-7835

[38] Zhou H, Pu S, Huo J, Cao W, Wang B, Li J. Facile corrosion synthesis and sintering of disperse pure tetragonal zirconia nanoparticles. Ceramics International. 2016;**42**:15005-15011. DOI: 10.1016/j.ceramint.2016.06.148

[39] Meetei SD, Singh SD. Hydrothermal synthesis and white light emission of cubic ZrO2: Eu3+ nanocrystals. Journal of Alloys and Compounds. 2014;**587**: 143-147. DOI: 10.1016/j.jallcom.2013.10.159

[40] Noh HJ, Seo DS, Kim H, Lee JK. Synthesis and crystallization of anisotropic shaped ZrO2 nanocrystalline powders by hydrothermal process. Materials Letters. 2003;**57**:2425-2431. DOI: 10.1016/S0167-577X(02)01248-X

[41] Nath S, Biswas A, Kaur PP, Sarma LS, Sur UK, Ankamwar BG. Synthesis of mesoporous nanocrystalline zirconia by surfactant-assisted hydrothermal approach. Journal of Nanoscience and Nanotechnology. 2018; **18**(8):5390-5396. DOI: doi.org/10.1166/jnn.2018.15362

[42] Gonell F, Portehault D, Julián-López B, Vallé K, Sanchez C, Corma A. One step microwave-assisted synthesis of nanocrystalline WOx–ZrO2 acid catalysts. Catalysis Science & Technology. 2016;**6**:8257-8267. DOI: 10.1039/x0xx00000x

[43] Majedi A, Davar F, Abbasi A, Ashrafi A. Modified sol–gel based

nanostructured zirconia thin film: Preparation, characterization,p and corrosion behavior. Journal of Inorganic and Organometallic Polymers and Materials. 2016;**26**:932-942. DOI: 10.1007/s10904-016-0394-7

[44] Huang Z, Han W, Feng Z, Qi J, Wu D, Wei N, et al. The effects of precipitants on co-precipitation synthesis of yttriastabilized zirconia nanocrystalline powders. Journal of Sol-Gel Science and Technology. 2019; **90**:359-368. DOI: 10.1007/s10971-019-04947-y

[45] Jaenicke S, Chuah GK, Raju V, Nie YT. Structural and morphological control in the preparation of high surface area zirconia. Catalysis Surveys from Asia. 2008;**12**:153-169. DOI: 10.1007/s10563-008-9048-2

[46] Huang HJ, Wang MC. The phase formation and stability of tetragonal ZrO2 prepared in a silica bath. Ceramics International. 2013;**39**:1729-1739. DOI: 10.1016/j.ceramint.2012.08.019

[47] Li X, Wang Y, Xiong X, Liu HB, Wu N, Liu R. Microstructure and synthesis mechanism of dysprosia-stabilized zirconia nanocrystals via chemical coprecipitation. Ceramics International. 2020;**46**(9): 13331-13341. DOI: 10.1016/j.ceramint.2020.02.112

[48] Dahanayake U, Karunaratne BS. Preparation of zircon ($ZrSiO_4$) ceramics via solid state sintering of ZrO_2 and SiO_2 and the effect of dopants on the zircon yield. Solid State Ionics. Advanced Materials for Emerging Technologies. 2006:146-153. DOI: 10.1142/9789812773104_0017

[49] Averu AC, Biel LV, Medeiros JLB, De Souza J. Comparative study of the diffusibility of a nickel alloy composed by Ni-Cr-Fe. International Journal of Engineering Research & Technology (IJERT). 2016;**5**(12):IJERTV5IS120261

[50] Aytimur A, Kocyigit S, Uslu I. Calcia stabilized ceria doped zirconia. Nanocrystalline ceramic. Journal of Inorganic and Organometallic Polymers. 2014;**24**:927-932. DOI: 10.1007/s10904-014-0064-6

[51] Ahmed T, Ren H, Noman M, Shahid M, Liu M, Ali MA, et al. Green synthesis and characterization of zirconium oxide nanoparticles by using a native Enterobacter sp. and its antifungal activity against bayberry twig blight disease pathogen Pestalotiopsis versicolor. Nano Impact. 2021;**21**:10028. DOI: 10.1016/j.impact.2020.100281

[52] Bansal V, Rautaray D, Ahmad A, Sastry M. Biosynthesis of zirconia nanoparticles using the fungus Fusarium oxysporum. Journal of Materials Chemistry. 2004;**1**(4):3303-3305. DOI: 10.1039/B407904C

[53] Al-Zaqri N, Muthuvel A, Jothibas M, Fahad AA, Alharthi A, Mohana V. Biosynthesis of zirconium oxide nanoparticles using Wrightia tinctoria leaf extract: Characterization, photocatalytic degradation and antibacterial activities. Inorganic Chemistry Communications B. 2021;**127**: 108507. DOI: 10.1016/j.inoche.2021.108507

[54] Kumari N, Sareen S, Verma M, Sharma S, Sharma A, SohalbHS MSK, et al. Zirconia-based nanomaterials: Recent developments in synthesis and applications. Nanoscale Advances. 2022; **4**:4210-4236. DOI: 10.1039/D2NA00367H

[55] Merten D, Broekaert JAC, Brandtc R, Norbert JJ. Analysis of ZrO2 powders by

microwave assisted digestion at high pressure and ICP atomic spectrometry. Journal of Analytical Atomic Spectrometry. 1999;**14**:1093-1098

[56] Chen S, Li J, Lu D, Zhu S. Study on the simultaneous determination of different volatile elements in high purity zirconium dioxide by in situ matrix removal and ETV-ICP-MS. Atomic Spectroscopy. 2016;**37**(1):13-18

[57] Degueldre C, Favarger PF, Bitea C. Zirconia colloid analysis by single particle inductively coupled plasma–mass spectrometry. Analytica Chimica Acta. 2004;**518**(1-2):137-142

[58] Stefanov P, Stoychev D, Stoycheva M, Ikonomov J, Ts M. XPS and SEM characterization of zirconia thin films prepared by electrochemical deposition. Material D. Chemistry. 2000; **30**(1):628-631

[59] Routray BCS, Mohanty R, Swain R, Sahu S, Mishra BR. Preparation and characterisation of zirconia nano-materials prepared from zircon minerals. Journal of The Institution of Engineers (India): Series E. 2021;**102**(1):87-95. DOI: 10.1007/s40034-020-00201-1

[60] Zhu LQ, Fang Q, He G1, Liu M, Zhang LD. Microstructure and optical properties of ultra-thin zirconia films prepared by nitrogen-assisted reactive magnetron sputtering. Nanotechnology. 2005;**16**:2865-2869. DOI: 10.1088/0957-4484/16/12/022

[61] Azdad Z, Marot L, Moser L, Steiner R, Meyer E. Valence band behavior of zirconium oxide, photoelectron and auger spectroscopy study. Scientific Reports. 2018;**8**:16251. DOI: 10.1038/s41598-018-34570-w

[62] Lackner P, Zou Z, Mayr S, Diebold U, Schmid M. Using photoelectron spectroscopy to observe oxygen spill over to zirconia. Physical Chemistry Chemical Physics. 2019;**32**: 1-9. DOI: 10.1039/C9CP03322J

[63] Barkadze G, Jeurgens PLH, Mittemeije EJ. Valence-band and chemical-state analyses of Zr and O in thermally grown thin zirconium-oxide films: An XPS study. Journal of Physical Chemistry C. 2011;**115**(40):19841-19848. DOI: 10.1021/jp206896m

[64] Deshmane VG, Adewuyi YG. Synthesis of thermally stable, high surface area, nanocrystalline mesoporous tetragonal zirconium dioxide (ZrO_2): Effects of different process parameters. Microporous and Mesoporous Materials. 2012;**148**(1): 88-100. DOI: 10.1016/j.micromeso.2011.07.012

[65] Sidhua GK, Kaushik AK, Ranac S, Bhansali S, Kumar R. Photoluminescence quenching of zirconia nanoparticle by surface modification. Applied Surface Sciences. 2015;**334**:216-221. DOI: 10.1016/j.apsusc.2014.10.036

[66] Ashkarran AA, Afshar SAA, Aghigh SM, kavianipour M. Photocatalytic activity of ZrO2 nanoparticles prepared by electrical arc discharge method in water. Polyhedron. 2010;**29**:1370-1374. DOI: 10.1016/j.poly.2010.01.003

[67] Fillit R, Homerin RP, Schafer RJ. Quantitative XRD analysis of zirconia-toughened alumina ceramics. Journal of Materials Science. 1987;**22**:3566-3570. DOI: 10.1007/BF01161460

[68] Wang J, Yin W, He X, Wang Q, Guo M, Chen S. Good biocompatibility and sintering properties of zirconia nanoparticles synthesized via vapor-phase hydrolysis. Scientific Reports. 2016;**16**:35020. DOI: 10.1038/srep35020

[69] Manoharan D, Loganathan A, Kurapati V, Nesamony VJ. Unique sharp photoluminescence of size-controlled sonochemically synthesized zirconia nanoparticles. Ultrasonics Sonochemistry. 2015;**23**:174-185. DOI: 10.1016/j.ultsonch.2014.10.004

[70] Harris DC, Bertolucci MD. Symmetry and Spectroscopy: An Introduction to Vibrational and Electronic Spectroscopy. 1st ed. New York, NY, USA: Oxford University Press; 1978

[71] Narayana DKL, Rao SV, Rao DN. Spectroscopic investigation of fs laser-induced defects in polymer and crystal media. Laser-Induced Damage in Optical Materials: Proc. of SPIE. 2012:8530

[72] Pezzotti G, Porporati AA. Raman spectroscopic analysis of phase-transformation and stress patterns in zirconia hip joints. Journal of Biomedical Optics. 2004;**9**(1):372-384. DOI: 10.1117/1.1647547

[73] Clarke DR, Adar F. Measurement of the crystallographically transformed zone produced by fracture in ceramics containing tetragonal zirconia. Journal of the American Ceramic Society. 1982;**65**:284-288

[74] Basahel S, Ali TT, Mostafa MMM, Katabathini N. Influence of crystal structure of nanosized ZrO2 on photocatalytic degradation of methyl orange. Nanoscale Research Letters. 2015;**10**(1):73. DOI: 10.1186/s11671-015-0780-z

[75] Shia F, Lib Y, Wanga H, Zhang Q. Fabrication of well-dispersive yttrium-stabilized cubic zirconia nanoparticles via vapor phase hydrolysis. Progress in Natural Science: Materials International. 2012;**22**(1):15-20. DOI: 10.1016/j.pnsc.2011.12.003

[76] Ada P, Falkowski P, Miazga A, Konopka K, Szafran M. Fabrication of ZrO2-Ti composites by slip casting method. Archives of Metallurgy and Materials. 2016;**61**(2B):1095-1100. DOI: 10.1515/amm-2016-0184

[77] Darmawan BA, Fisher JG, Trung DT, Sakthiabirami K, Park S. Two-step sintering of partially stabilized zirconia for applications in ceramic crowns. Materials. 2020;**13**(8):1857. DOI: 10.3390/ma13081857

[78] Anjum DH. Characterization of nanomaterials with transmission electron microscopy. IOP Conference Series: Materials Science and Engineering. 2016:146

[79] Siddiqui MRH, Al-Wassil AI, Al-Otaibi AM, Mahfouz RM. Effects of precursor on the morphology and size of ZrO2 nanoparticles, synthesized by sol-gel method in non-aqueous medium. Materials Research. 2012;**15**(6):986-989. DOI: 10.1590/S1516-14392012005000128

[80] Liu J, Mir AH, He G, Danaie M, Hinks J, Donnelly S, et al. In-situ TEM study of irradiation-induced damage mechanisms in monoclinic-ZrO2. Acta Materialia. 2020;**199**:429-442

[81] Salih SI, Salih WB, Hamad HS. Preparation and investigation of some mechanical properties of unsaturated polyester resin reinforced by hybrid nanoparticles. The Iraqi Journal For Mechanical And Material Engineering. 2018;**18**(2):331-338

[82] Lee M, Son JS, Kim K, Kwon T. Improved resin–zirconia bonding by room temperature hydrofluoric acid etching. Materials. 2015;**8**:850-866. DOI: 10.3390/ma8030850

Chapter 4

Processing of Zirconia

Ghassan Albarghouti and Sojood Mahmoud Farhan Darwish

Abstract

This chapter starts with an introduction, including zircon, zircon structure, zirconia ceramics and their structures as they exhibit a distinct crystal structure at various temperatures while maintaining the same chemical composition. Then the properties of zirconia including mechanical, tribological, and electrical properties and thermal conductivity, were discussed. Zirconia's properties offer excellent resistance to corrosion and chemicals. When compared to other high-tech ceramic materials, zirconia is very robust at room temperature. The processing of zirconia was explained, starting with the purification of zirconium compounds followed by powder processing of zirconia, colloidal processing of zirconia, additive manufacturing, and zirconia treatments. Zirconia ceramics are processed from their raw ingredients in order to modify and enhance their physical and chemical characteristics and make them more suitable for use in future processes and in the production of finished goods. Various analysis methods of zirconia, including powder processing and sintering, microstructural analysis, phase characterization, mechanical testing, and tribological characterization were discussed. The last section in the chapter discusses the nano zirconia (ZrO_2 nanoparticles). In addition to photocatalytic and piezoelectric uses, dental and optical coatings, nano zirconia has additionally been employed as a catalyst in a variety of organic interactions. It appears in the cubic, tetragonal, and monoclinic structural phases.

Keywords: zirconia, structure, properties, processing, analysis, 3D printing

1. Introduction

1.1 Zircon

Zircon, which is frequently found in highly silicic stones and contains significant quantities of certain trace elements, is regarded as a resistant auxiliary mineral. It is also a helpful instrument for tracking melt compositional alterations throughout magma development [1]. The quantity of Zr in the principal minerals is often negligible in comparison to that housed by zircon, and there is disagreement on the presence of zircon at boundaries of grains vs. the amount hidden as inclusions within different minerals [2]. Zirconium silicate is another name for the mineral zircon ($ZrSiO_4$). It results from the extraction and processing of historical heavy mineral sand and is most typically found in coastal placers or dune deposits across the world [3]. Zirconium silicates are abundant in essence, along with presence under hydrothermal circumstances (between 300 and 550°C) which seems to have received a lot of interest [4].

The global output of zircon is around 1.1 million t/a, with the majority of it coming from Australia and South Africa. The most important use of zircon is represented by ceramics sector [3]. Zircon may be treated to make zirconia by melting the sand at extremely high temperatures to yield molten zirconium oxide (ZrO_2) [5].

It is debatable how zircons behave in suprasolidus metamorphic environments. Experimental findings suggest zircon growth occurs throughout prograde metamorphism in melt-bearing systems, but computational modeling only anticipates zircon development upon cooling and melt crystallization. Zircon in high-pressure zones and elevated temperatures metamorphic rocks is a probable by-product of inconsistent melting events. However, knowledge of how zirconium, a crucial structural component of zircon, is distributed among the numerous products and reactants that occur during incongruent partial melting is still limited. A fresh source of Zr may become accessible to zircon formation after heating under suprasolidus under certain circumstances, or Zr may redistribute from already-existing zircon [2].

1.2 Zircon structure

Zircon is a mineral that is an orthosilicate accessory. Its chemical formula is $ZrSiO_4$ ($Z = 4$) with the space group $I4_1/amd$. The zircon structure is made up of chains of consecutive, edge-shared SiO_4 and ZrO_8 polyhedra that pass along the c axis. The structure is rather wide; it incorporates "channels" of evidently vacant sixfold sites that run in corresponding direction to the {001} orientation and can accommodate big amount of non-formula components [6].

The edges of the ZrO_8 triangular dodecahedra are coupled in a zigzag manner across the {100} dimensions. The structure shown in **Figure 1** accounts for the most prevalent three-dimensional behavior of zircon as well as the uncertain cleavage along {110}. Natural zircon incorporates non-formula elements at all times [7]. On the other hand, large cations (U, Th, Y, lanthanides, Hf) often replace Zr in the eightfold dodecahedral position, however, smaller cations may be placed in the sixfold location [8].

1.3 Zirconia ceramics

For several reasons, zirconia-based ceramics are the most researched and demanding materials. Zirconia (zirconium dioxide, ZrO_2), popularly known as "ceramic steel", offers outstanding hardness, strength, and fatigue resistance, as well as good wear characteristics and biocompatibility [9].

Zirconium oxide, silicon nitride, and aluminum oxide are a few frequently employed ceramics that have exceptional chemical and physical qualities including high hardness, high strength, high toughness, and corrosion resistance [1]. Zirconium oxide, on the other hand, outperforms silicon nitride and aluminum oxide in terms of abrasion resistance, toughness, and service temperature [10].

Zirconia is a significantly highly versatile transition metal oxide frequently applied in ceramics, electrolytes for solid-oxide fuel cells and semiconductor technology [11]. The finest qualities come from zirconia stabilized with Y_2O_3. A crystallographic change on a ZrO_2 surface limits fracture formation once is subjected to stress [12].

Baddeleyite, a naturally existing zirconia, is found in igneous rocks such as carbonatite. South Africa was the major producer of baddeleyite until 2002, but these plants have since closed, leaving the Kola Peninsula area of the Russian Federation as the only commercial source of this mineral. The majority of zirconia presently

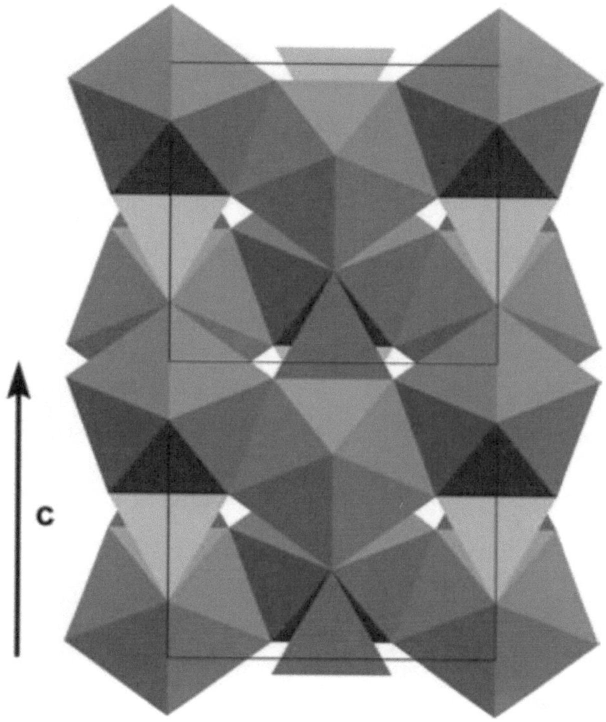

Figure 1.
Two-unit cells shown in the polyhedral depiction of the zircon structure [7].

accessible is made using zircon (34,000 t/a), with a relatively minor natural source by the Russian Federation (6000 t/a) [3].

Zirconia ceramics are traditionally processed using turning, milling, and grinding techniques. These techniques have significant drawbacks, including inadequate productivity and inevitable manufacturing damage, which make it difficult to achieve sufficient quality of the surface. But sandblasting and acid etching have a tendency to pollute the component's surfaces or alter their physical and chemical characteristics. In order to get an immaculate finish and undo surface and subsurface damage from earlier preparation, polishing can significantly enhance the surface quality [10].

1.4 Zirconia structure

There are three different types of zirconia crystal structures: Cubic, tetragonal, and monoclinic. It is possible to achieve high structural stabilization by combining ZrO_2 and associating to other metallic oxides like ZrO_2—MgO, ZrO_2—CaO, or ZrO_2—Y_2O_3 [13]. However, recent research emerged to be more concentrated on zirconia—yttria ceramics, which can be recognized by their fine-grained in the micro range classified as Tetragonal Zirconia Polycrystals (TZP) [14].

At high temperatures (>2370°C), a cubic zirconia also known as (c-ZrO_2) fluorite crystal configuration with a unit cell of face-centered cubic (fcc) and a (space group Fm-3m) form. A numerous atom of Zr is linked to eight oxygens to create the fcc cube and the oxygen atoms are arranged along the cube's diagonals [15]. Furthermore, the fluorite structures of the monoclinic and tetragonal are deformed as seen in **Figure 2**.

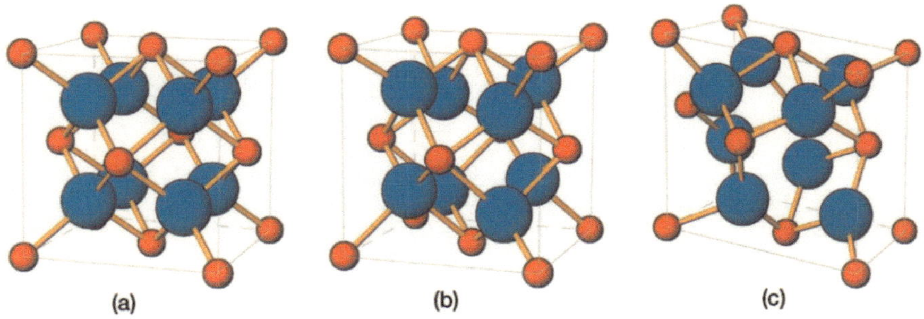

Figure 2.
The three ZrO_2 polymorphs are shown in a schematic in the following order: (a) cubic, (b) tetragonal, and (c) monoclinic [16].

At intermediate temperatures (1200–2370°C), the structure is transferred to tetragonal ($P4_2/nmc$), while at low temperatures (<950°C), the structure is monoclinic ($P2_1/c$) [16]. Throughout thermal cycling, undoped zirconia reveals the following phase transitions [17]:

$$\text{m-ZrO}_2 \underset{950^\circ\text{C}}{\overset{1170^\circ\text{C}}{\rightleftharpoons}} \text{t-ZrO}_2 \overset{2370^\circ\text{C}}{\rightleftharpoons} \text{c-ZrO}_2 \overset{2650^\circ\text{C}}{\rightleftharpoons} \text{Liquid} \quad (1)$$

2. Properties of zirconia

In relation to their usage, this section describes the many varieties and characteristics of zirconia materials. Due to the following qualities, zirconia has recently been the subject of substantial inquiry in the scientific literature:

2.1 Mechanical properties

Recent studies showed that mechanical parameters including hardness, grain size, and fracture toughness, as seen in **Table 1** have an important bearing on slide wear resistance of bulk ceramics [18]. Zirconia has mechanical properties that are on par with stainless steels. Its traction resistance might be in the 900–1200 MPa range, and it has a 2000 MPa compression tolerance [19]. 3Y-TZP dentistry zirconia incorporates alumina (Al_2O_3) as a strain-hardening promoter, resulting in great opacity and exceptional mechanical qualities [20].

Although the sintering of ZrO_2 stabilized with Y_2O_3 is substantially more complex, this is the primary type of zirconia recognized for present medical usage. Surface treatments have the potential to alter the physical characteristics of zirconia. Furthermore, prolonged wetness contact may have a negative impact on its assets. The term for this is zirconia aging. Surface grinding can also diminish toughness. The mechanical characteristics of zirconia might deteriorate as it ages. Mechanical forces and moisture exposure are essential for speeding up this process [20].

Due to transition toughening, ZrO_2 has the ability to achieve extremely high toughness. It is regarded as a strong contender for numerous cutting tool applications as well as a biomedical material. Nevertheless, the intrinsic fragility of these substances remains the primary bottleneck. This issue has been addressed by developing suitable composites. Novel materials, including carbon nanotubes and fiber-like structures, have been produced in recent years. The fibrous character of the

Property	Minimum value (S.I.)	Maximum value (S.I.)	Units (S. I.)	Minimum value (Imp.)	Maximum value (Imp.)	Units (Imp.)
Atomic volume (average)	0.02	0.021	$m^2\ kmol^{-1}$	1220.47	1281.5	$In^3\ kmol^{-1}$
Density	5	6.15	$Mg\ m^{-3}$	312.14	383,932	$lb\ ft^{-3}$
Energy content	200	300	$MJ\ kg^{-1}$	21667.7	32501.6	$kcal\ lb^{-1}$
Bulk modulus	72.3	212	GPa	10,4862	30,748	10^6 psi
Compressive strength	1200	$200	MPa	174,045	754,196	Ksi
Elastic limit	115	711	MPa	16.6793	103.122	Ksi
Endurance limit	107	640	MPa	15,519	92.8241	Ksi
Fracture toughness	1	8	$MPa\ m^{1/2}$	0.910047	7.28037	$ksi\ in^{3/2}$
Hardness	5500	15,750	MPa	797,708	2284.35	Ksi
Loss coefficient	0.0005	0.001		0.0005	0.001	Null
Modulus of rupture	177	1000	MPa	25.6717	145,038	Ksi
Poisson's ratio	0.22	0.32		0.22	0.32	Null
Shear modulus	53.4	86.4	GPa	7.74501	12.5313	10^6 psi
Tensile strength	115	711	MPa	16.6793	103,122	Ksi
Young's modulus	100	250	GPa	14.5038	36.2594	10^6 psi
Latent hear of fusion	700	820	$kJ\ kg^{-1}$	300,944	352,535	$BTU\ lb^{-1}$
Maximum service temperature	1248	2522	K	1786.73	4079.93	°F
Melting point	2823	2973	K	4621.73	4891.73	°F
Minimum service temperature	0	0	K	−459.67	−459.67	°F
Specific heat	420	540	$J\ kg^{-1}\ K^{-1}$	0.32502	0.417883	kJ/kg °K
Thermal conductivity	1.7	2.7	$W\ m^{-1}\ K^{-1}$	3.18246	5.05449	$W\ K^{-1}\ m^{-1}$
Thermal expansion	2.3	12.2	$10^{-6}\ K^{-1}$	4.14	21.96	$10^{-6}\ °F^{-1}$
Breakdown potential	4	6	$MV\ m^{-1}$	101.6	152.4	V per mil
Dielectric constant	10	23	—	10	23	Null
Resistivity	3.16×10^{13}	3.16×10^{18}	10^{-8} ohm m	3.16×10^{13}	3.16×10^{18}	10^{-8} ohm m

Table 1.
Thermal, electrical, and mechanical properties of zirconia [18].

toughening components may result in increased fracture toughness. Furthermore, compared to the other two non-metallic additions, carbon's electrical conductivity allows for additional electrical qualities to be helpful [21].

Using Eq. (2), it is possible to conduct a three-point bending test and a biaxial flexure test to assess flexural capacity and can be examined using a universal testing device (Autograph AG-X).

$$\sigma = \frac{-0.2387P(X - Y)}{b^2} \tag{2}$$

where the center's highest tensile stress is σ (in MPa), the total load producing the rupture is P (in N), and the cross-section of the test piece at the fracture point is b (in mm) [22].

Meanwhile, fracture toughness (K_{IC}) test can be performed using the single-edge pre-cracked beam technique and can be measured by applying Eq. (3). Where **B** represents the material length (in m), **W** represents the sample width (in m), and **a** represents the pre-crack length; **P** represents the breaking force (N), and **S** represents the support span (m) [22].

$$K_{IC} = \left(\frac{P \times S}{B \times W^{3/2}}\right) \times \left\{\frac{3}{2}\left(\frac{a}{W}\right)^{\frac{1}{2}} \times Y\left(\frac{a}{W}\right)\right\} \tag{3}$$

2.2 Tribological properties

Zirconia ceramics have been extensively researched as a wear resistant material for engineering fields throughout the last few years. The kind of Hall-Petch law governs the connection between wear resistance and grain size in tetragonal zirconia ceramics. As a result, it is reasonable to predict that nano-structured zirconia coatings will have more wear resistance than standard zirconia. Friction and wear tests may sometimes be performed on an MM-200 wear tester's block-on-ring configuration. As seen in **Figure 3**, the stationary zirconia-coated blocks are

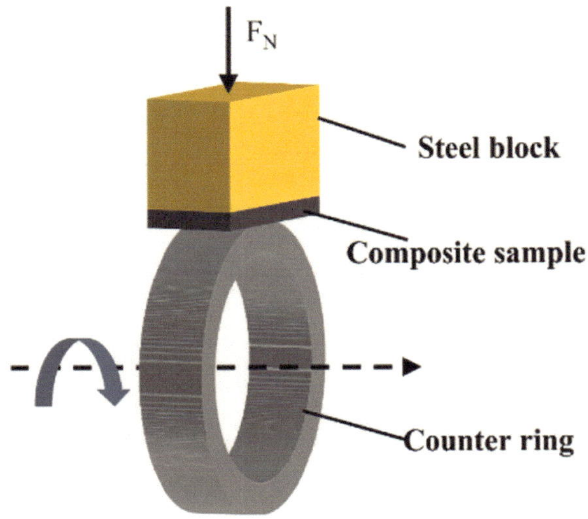

Figure 3.
The block on ring test [23].

pressed up against a moving stainless steel. The tester provides the friction coefficient immediately. Additionally, wear frequencies can be calculated by dividing the wear mass loss by the force implemented, sliding distance, and material density [24].

Throughout wear testing, the surface morphology, particularly big pores, may result in excessive stress concentration and fracture development, resulting in poor resistance. Additionally, it is shown that, as evidenced by Eq. (4), the wear rate of polycrystalline ceramics matched an exponential function. In which **$1/K_{0w}$** represents the TZP substance wear resistance with no pores (in N.m/mm^3) and **P** represents porosity.

$$\frac{1}{k_w} = \frac{1}{k_{0w}} exp\,(-23\,P) \tag{4}$$

Eq. (4) above illustrates that decreasing the porosity of zirconia ceramics resulted in enhanced wear resistance. The existence of gaps reduces the probability of grain boundary sliding, which delays or prevents the occurrence of plastic deflection [24].

2.3 Electrical properties and thermal conductivity

Zirconia additionally has a high degree of flexibility, a superior conductivity, a poor thermal conductivity, and a strong corrosion resistance as seen in **Table 1** [18]. Because of that, zirconia is utilized in many applications at various high temperatures, such as fuel cells and thermal barrier coatings [18]. Monoclinic zirconia was demonstrated at 1000°C to be an amphoteric semiconductor in a prior work. When oxygen pressures are high (10^{-6} to a value of 1 atm), the most common defects are fully ionized zirconium holes, which produced exceptionally well, implying charge transfer via a thermally induced hopping mechanism [25].

On the other hand, at ambient temperature, zirconia in its monoclinic phase has a thermal conductivity of 7.2 $Wm^{-1}\,K^{-1}$. However, at high temperature (1100°C) zirconia begins to have very poor heat conductivity (1.2–2.6 $Wm^{-1}\,K^{-1}$), making it an ideal candidate for thermal barrier (TBC) coatings [26]. However, the volume variation (5%) caused by phase transformation at elevated heat as well as during cooling the apparatus limits the use of zirconia in aggressive environments. The crumbling of the zirconia-based components is caused by phase transformation and volume change [27].

Yet, when the temperature rises, the phase transition (**Figure 4**) occurs, resulting in structural failure and fractures in the coat. This would be prevented by enriching the zirconia matrix with yttria, which stabilizes the zirconia at high temperatures. Some Zr^{4+} cations are substituted by Y^{3+} during this stabilizing procedure [29]. To preserve charge balance, one oxygen gap is formed for every two replacing Y^{3+} cations. Because of the presence of voids, YSZ is valuable not only for TBCs but also as an electrolyte in oxygen sensors and solid-oxide fuel cells (SOFC) [30].

This kind of stabilization not only assists in preventing phase change but also reduces the thermal conductivity of YSZ, for example, from 1.42 W/m·K (at ambient temperature) to 1.35 $Wm^{-1}\,K^{-1}$at 1200°C since c-phase zirconia exhibits less thermal conductivity than m-zirconia [27].

Zirconia is also a metal oxide which is composed of a weak base and a weak acid and is one of the ceramic semiconductors. Because of its crystalline structure, it can be

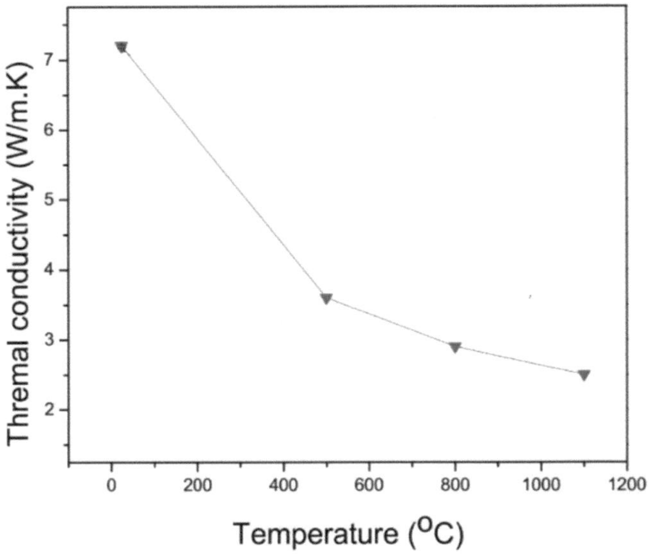

Figure 4.
Thermal conductivity of ZrO_2 [28].

an insulator utilized as an n-type semiconductor or a high-resistance ceramic. The forms of the nanocrystals determine the potential uses for them. Oxygen sensors, fuel cell electrolytes, and gate dielectrics have all been employed with spherical ZrO_2 [31].

3. Processing of zirconia

This section discusses the purification process of zircon, followed by stating various methods for zirconia processing (thermal dissociation, chlorination, additive manufacturing, and colloidal processing) depending on the characteristics asked for and the purpose or the industry that asks for a zirconia supply.

3.1 Purification of zirconium compounds

For usage in the electronics sector and the creation of partly stabilized zirconia, highly pure zirconia is commonly used today. A large portion of the world's zirconia supply found naturally as $ZrSiO_4$ [32].

By caustic fritting of zircon sand, the zirconium nitrate solution is obtained. Caustic fritting was accomplished with the aid of magnesium hydroxide carbonate and magnesium oxide. These additives reduce the quantity of soluble silica in nitric acid by less than 1000 ppm. With no additional ingredients, caustic fritting produces some nitric acid-soluble silica, which is eliminated by dehydrating with sulfuric acid [32].

Other procedures currently are utilized to extract zirconia from zircon generally by including heat or chemical disintegration of zircon to provide zirconia and silica. But even so, the resulting zirconia needs to be purified through additional chemical processing. A financially appealing method of producing zirconia is the thermal dissociation of zircon, which results in a combination of zirconia and silica. The dissociated zircon result can be leached either with a strong base, like sodium hydroxide

(NaOH), to breakdown the silica and end up leaving zirconia or with a strong acid, like sulfuric acid (H_2SO_4), to solubilize the zirconia and produce a zirconium salt, leaving non-soluble silica. Zirconia may be created using the caustic-leach technique with a purity of up to 99.5%. Still, it must undergo subsequent chemical treatment, usually including dissolving the zirconia in acid to attain greater purity zirconia [32].

In reality, however, acid-leaching precipitates are gel-like, challenging to filter, and enclose numerous contaminants. Because of this, the basic sulfate precipitation using zirconium sulfate solution is favored, even though it is challenging to regulate the circumstances. To precipitate the basic sulfate, sulfuric acid or a sulfate salt is frequently added after the zirconium sulfate is converted to zirconium oxychloride. It has been discovered that creating zirconium compounds with good quality, including zirconia, is feasible by precipitating zirconium from zirconium sulfate solution at a low pH [32].

It is desired to utilize a zirconium sulfate solution with a zirconium level of more than 75 g/L and a sulfate content of more than 180 g/L. The ammonia source is introduced to the aqueous zirconium sulfate solution throughout the procedure of the present invention till the solution's pH value is between 0.1 and 2.5. Whenever the pH is in the interval of 1.0 to 2.0, it is preferable to stop adding the ammonia source. The zirconium compositions created using the procedure of the present invention are white, compressible, unrestricted, finely dispersed substances after washing and drying [32].

3.2 Chlorination

Chlorinating impure zirconia in the presence of carbon is among the best ways to create highly pure zirconia. To purge contaminants from the $ZrCl_4$, partial solidification and sublimation procedures are performed in the range of temperature between 623 and 673 K. Additionally, highly pure ZrO_2 is created by the interaction of pure solid $ZrCl_4$ with water steam. Another crucial stage in the industrial scale of titanium and zirconium is chlorination in the presence of carbon. In the temperature range of 1000 K–1100 K, chlorine and ZrO_2 do not really thermodynamically interact, and a reducing agent like carbon is necessary to complete the process. Consequently, the process is known as carbochlorination [33].

By determining the variation in the free energy of the different reactions, it is possible to know why a reducing agent is required. By creating a low oxygen potential environment, the existence of carbon atoms lowers the inclination of the production of oxides and stimulates the creation of chlorides. It has been discovered that between 1075 K and 1275 K, the following reaction ($CO_2 + C = 2CO$) moves more slowly than during the chlorination process of ZrO_2 with a ΔG of -4.6 kJ mol^{-1} at 1000 K to a -22.0 kJ mol^{-1} at 1100 K [34]. On the other hand, thermodynamically speaking, reaction 1 cannot take place at 1100K but other reactions may. It's possible that reaction ($ZrO_2 + C + 2Cl_2 = ZrCl_4 + CO_2$) with a ΔG of -248.3 kJ mol^{-1} and reaction ($CO_2 + C = 2CO$) will produce reaction ($ZrO_2 + 2C + 2Cl_2 = ZrCl_4 + 2CO$) with a ΔG of -270.4 kJ mol^{-1}.

3.3 Colloidal processing of zirconia

During lower sintering temperatures, high-density zirconia structures with decently crystal frames and improved mechanical characteristics may be created using fine nanocrystalline 3Y-TZP powders. However, throughout most of the production of the 3Y-TZP nanocrystalline initial crystals, significant agglomeration takes place [35]. The outer surface of the colloidal matter is decreased by agglomerates, which consist of networks of particles linked by van der Waals forces [36]. Ultimately, accumulation could lead to the development of undesirable variabilities, including defects, fractures, and porous surfaces with big granules, which can impair the mechanical capabilities of zirconia structures. As a result, attempts must be considered to prevent agglomeration due to consolidation. Powders are merged in greenish structures with the necessary shapes using various consolidation techniques, such as solid free-form manufacturing, wet slip casting, dry isostatic and uniaxial pressing. A slip casting is a common colloidal procedure that creates structures with a significant density and few harmful variabilities. By adjusting the intensity of the attracting and repelling interparticle interactions between zirconia particles, steady solutions with high particle dispersion may be created, which are necessary for reducing the production of agglomerates [37]. An electric double surface is produced by generating comparable charges on the interfaces of suspended particles of sufficient size [36].

CAD/CAM, which denotes computer-aided engineering and production, is one way to create zirconia copings and frames. An automatic manufacturing process using a software power tool follows the editing of a 3D model on a screen in the CAD/CAM manufacture of zirconia. There are two methods for creating CAD/CAM zirconia foundations: pre-sintered blanks can be "soft-machined" or completely sintered blanks can be "hard-machined". These foundations have undergone high-quality procedures to produce homogeneous material structures with little to no cavities, defects, or fractures. A binder is blended with partly stabilized zircon powder (silicone oxide 0.02%, yttrium oxide >4%, hafnium oxide >1%, zirconium dioxide 96%, aluminum oxide 1%) to create zirconia [38].

Well-dispersed zirconia solution has just been subjected to direct coagulation casting (DCC) and gel casting are two innovative methods for treating colloidal materials to produce greenish particles [39]. DCC focuses on creating ceramic components closer to their final form from suspensions of ceramic particles in concentration solutions. It combines the DLVO concept with colloidal ceramic formation. The saturated suspension is disrupted by the decomposition generating base, electrolyte, and acid, whether to boost the ionic strength or to let the pH reach the ceramic powder isoelectric point framework [35, 36].

3.4 Powder processing of zirconia

Because of their acceptable mechanical qualities, minimal neutron absorption cross-section, and superior corrosion tolerance, zirconium that concentrates on alloys has been employed extensively in industrial power stations.

Nevertheless, because zirconium has a great melting point (>1800°C) and a significant attraction for H, O, and other contaminating atoms, it is challenging to produce zirconium workpieces using standard methods. AM will make it easier to produce complex parts that are employed in a variety of industrial applications. The quality of the finished products, however, is heavily dependent on the raw powders. When used in the AM process, powders with desirable morphological properties display improved rheological properties and more optimum particle packing. Research on producing sphere powder from commercial products like Ti64 has advanced significantly. Spherical metal components have been created using various techniques, such as Electrode Induction Melting Gas Atomization, Plasma Atomization, Plasma Rotating Electrode Process, Plasma Spheroidization (most flexible approach), and Vacuum Induction Melting Inert Gas Atomization [40].

3.5 Additive manufacturing (AM)

3D printing technology commonly referred to as additive manufacturing techniques, opens up new possibilities for material shaping by increasing design flexibility, accelerating production, and minimizing waste and expense. The revolutionary possibilities of using these technologies to produce technical ceramics have been widely acknowledged, despite the ceramic industry's slower adoption of 3D printing technology than the polymer and metal industries. Stereolithography (SLA), one of the several 3D printing technologies currently accessible, offers advantages in the areas of printing resolution and precision, which—along with the exceptional efficiency of exterior coating—make it the ideal suitable AM technology for constructing ceramics [41]. The greatest range for achieving uniform mixing, according to an investigation by Jang et al. on the synthesis of zirconia employing DLP additive manufacturing, was 58 vol% of zirconia by volume. The 3-point bending durability improves as the volume fraction rises. The research did note that the viscosity rose quickly to 56 vol% [42]. SLA technique is built on the layer-by-layer polymerization of a liquid photocurable monomer incorporating ceramic particles; more specifically, it uses a beam of UV laser that shifts from one location to another while tracking the photo-polymerized pattern [43].

A near fabrication path to the manufacture of progressed ceramics with extremely complicated geometries is provided by additive manufacturing (AM). It does not involve the original building of a mold, making it a quick-response solution for engineers and designers to create new items [44].

A possible method for creating intricately shaped ceramic components with high accuracy is digital light processing (DLP) [43]. A different type of SLA utilizes a UV projector as the light source for treating an entire resin layer by layer, improving the speed of the process. Due to the variety of utilizations for which this material can be applied, including restorative dentistry and frameworks, fuel cells for monolithic support, tooling and blades, and other precision parts for various applications, including mechanical and thermal ones, 3D-printed zirconia is becoming more and more popular among technical ceramics [44].

DLP, which has excellent printing resolution, a high interfacial polish, and a rapid building process, is becoming increasingly popular for manufacturing ceramic components. However, compared to traditional ceramic processing, DLP is still in its developing phase [44].

Making completely dense, defect-free ceramics with characteristics equal to those of materials treated using traditional technologies is considered a difficult task. This requires the supervision of numerous manufacturing steps, including printing optimization techniques, the slurry preparation, and the sintering heating cycles. Nevertheless, even though DLP and SLA are established for zirconia 3D printing technologies, most articles briefly discuss the mechanical characteristics of the resulting materials [43] within the beginning, zirconia green entities are created for the specimen preparation using a heavily zirconia-loaded slurry on a DLP 3D printer utilizing a UV light with a wavelength of 405 nm. In **Figure 5**, a diagram of the DLP printing principles is displayed. For all builds, the depth can be adjusted at 25 μm. For further characterization and assessment, a variety of zirconia green bars, disks, and parts are acquired.

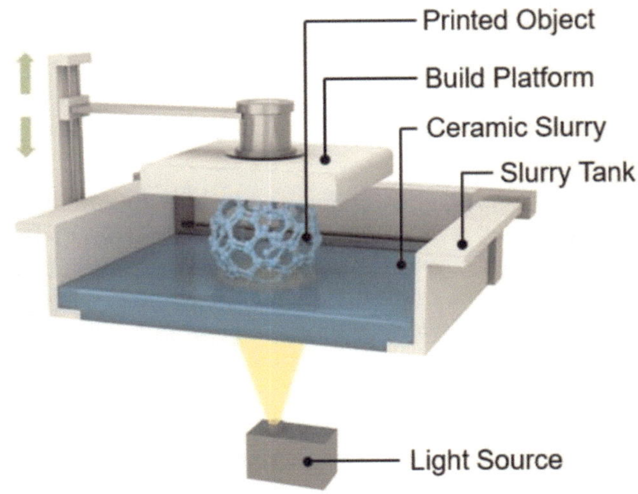

Figure 5.
Diagram illustrating the DLP printing principles [39].

The extra slurry is then washed off the artificial green parts in an ultrasonic bath. The debinding and sintering then performed in a tubular furnace with an air flow [44].

3.6 Thermal dissociation/calcination

Calcination is the practice of heating a substance at 400, 600 and 800°C without allowing it to fuse to affect changes in its physical or chemical composition [45]. Electrospinning, a newly improved combination of the ceramic precursor zirconium acetylacetonate and the bond polymer polyacrylonitrile, may be utilized to create zirconia incorporating nanofibers. It is discovered that the steps that follow the ceramic precursor conversion occur simultaneously with the transformation of electrospun zirconium acetylacetonate/polyacrylonitrile fibers into zirconia nanofibers:

$$Zr(C_5H_7O_2)_4 \rightarrow Zr(OH)(CH_3COO)_3 \rightarrow ZrO(CH_3COO)_2 \rightarrow t - ZrO_2 \rightarrow t - ZrO_2 + m - ZrO_2 \rightarrow m - ZrO_2. \quad (5)$$

Fibers are heated in two stages; first, they are heated to a temperature of 500°C with a heat rate of 1°C/min, and subsequently to a specified temperature with a heat rate of 5°C/min, calcining them at various temperatures between 500 and 1300°C for 1 hour. A low-temperature profile was adopted for the first annealing stage to ensure the sensitive removal of the ceramic precursor's breakdown products and binding polymer to minimize fiber breaking [46].

Additionally, furnaces for manufacturing glassware have often been built using and repaired with zircon refractories. Its breakdown tendency is a crucial zircon property that might affect furnace and glass performance. A decrease in the proportion of aluminum, titanium, iron, and alkali is necessary to lessen the rate of zircon deterioration [47].

Generally, the below process may be used to describe zircon dissociation:

$$ZrSiO_4 \rightarrow 3\ ZrO_2 + SiO_2 \tag{6}$$

where zircon is transformed into zirconia and silica, often by thermal methods at a temperature of around 1676°C. Zirconia acquires a tetragonal shape at 1173°C, whereas underneath this temperature, it exists as a monoclinic phase [47].

Zirconia precipitates are calcined at 800°C for 2 hours (h) in the air, which changes the monoclinic crystal structure and improves the crystallinity as well as the dimensions of the precipitate. Additionally, it was noted that zirconia with a cubic and tetragonal structure undergoes a partial transformation from its monoclinic state. In additional research, acidic and ammoniacal zirconium chloride solutions in different concentrations were used to create nanoscale zirconia precipitates at temps between 110 and 150°C for times ranging from 1 to 4 hours [48].

3.7 Zirconia treatments

Surface treatment to form a strong chemical connection with such composite resin is one method for successfully mending chemically inert zirconia foundations. Consequently, the effectiveness of dental restorations is determined by the adhesiveness of zirconia and composite resin.

It is extremely hard to chemically or physically change the zirconia ceramic surfaces because of its superior stiffness and chemical inertness to avoid corrosion [49]. In addition, unlike other types of ceramics, such as glass, zirconia is not etchable, making it difficult to carry out the adhesion processes [50]. According to earlier research, sandblasting significantly enhanced the ceramic surface's wettability, surface energy, and roughness. Because of this, the resin cement is retained by micro-interlocking due to the bonding area's enhancement [49].

However, implant technology has been working hard to change zirconia in terms of morphological and bioactive features essential for suitable cell interaction and differentiation throughout most of the neighboring bone healing. Numerous physicochemical techniques have been utilized to improve zirconia properties, such as PVD, grit blasting, laser treatment, PVD, micro-machining, and CVD [51].

Although it has been asserted that zirconia pretreatment techniques like tribochemical silica coating, sol-gel processes, silicon nitride hydrolysis, and vapor-phase deposition technique increase resin adherence to zirconia, the long-term bond performance of tribochemical silica coating is in doubt. Condensed silanol layers produced by the sol-gel technique and hydrolyzed from tetraethyl orthosilicate are prone to cracking when heated, and coatings produced by silicon nitride hydrolysis or vapor-phase deposition require complex equipment and take a long time, making them unsuitable for use in clinical settings. With 5% HF inscription for 90–120 seconds at a pressure of 0.3 Mpa, a new glass ceramic spray deposition (GCSD) approach has recently been described to increase the zirconia bond strength. It is possible to create an extremely thin layer that does not alter the zirconia's physical characteristics by spraying glass-ceramic powders over zirconia surfaces which are subsequently sintered [52].

3.8 Powder sintering

During sintering, powder components are condensed and go under consolidation at a high temperature but still below their melting temperature. A particle substance is transformed into a rigid, compact design. Somewhere at macro scales, changes in geometry, dimension changes (shrinkage), and density variation may be seen, as well as changes in mechanical qualities like mechanical strength (as a result of shrinkage) [53].

To achieve a greater compactness and performance, more machined and sintering operations are required [44]. However, these existing techniques methods have a variety of drawbacks, including the inability to produce highly complex geometry because of the common usage of molds, their increased price, and their prolonged processing times. Furthermore, machining ceramic materials is challenging due to their extraordinary hardness and brittleness. Additionally, flaws and unwanted shrinkages may also be produced in the ceramic parts [54].

DLP-produced sintered ceramic components may have shrinkage problems. An effective way to address them is to scale the CAD/CAM model appropriately for each axis before printing to account for shrinks [44]. Through DLP, a single layer of material is first made, and by repeating this process, a body with three dimensions is developed. The three-dimensional cross-linked polymer that makes up the green body has ceramic fragments caught inside it. The components are then cleaned, debonded, and sintered to create the final thick ceramic part. This technique makes it possible to create intricate three-dimensional ceramic structures with excellent precision and creative freedom [55].

The microstructure changes throughout sintering, and this development is characterized by a rise in grain compaction and rearrangement and cohesive bonds among granules develop in the early stages. Because of the mass transmission, the bonds between particles expand as the sintering process continues. Surface and grain boundary diffusion are often the primary mass transport modes in sintering. Surface tension and bonds' stresses cause particle attraction, which causes the system to contract. The decrease and occasionally the actual eradication of material porosity is a result of neck development and shrinking [53].

Owing to the possibility of the ceramic material's cracking or breaking, thin restorations for minimally invasive dentistry might be challenging to create employing the subtractive method. In addition, subtractive manufacturing usually results in significant cutting tool usage and significant volumes of generated waste (zirconia powder) after milling. Additionally, zirconia's resilience and lifetime are improved by surface treatment and adhesive bonding with primers [55].

4. Analysis methods

4.1 Microstructural analysis

For microstructural characterization, Raman microanalysis can be demonstrated. On the other hand, for studying the chemical composition, high vacuum X-ray energy dispersive microanalysis, X-ray photoelectron spectroscopy (XPS) are usually used. XPS is accelerated under a voltage of 14 kV, emission of current of o.2 A, pressure of 10^{-9} mbar, energy resolution of 0.30 eV, maximum power of 2.8 kW, sampling area of 6×0.5 mm sampling area, and around 90° take-off angle. Finally, to charge compensation, an electron flood gun works at 4 eV. All spectra can be shown and recorded on the binding energy with an excellent resolution of 150 eV pass energy [56].

In order to recognize and define the pattern of the tetragonal (which can be demonstrated by 145 and 262 cm^{-1} Raman shift) and monoclinic (by 180 and 190 cm^{-1} Raman shift) ZrO_2 phases just at the surface area, then the cervical collar that is prepared for XRD investigation is examined. The following requirements are used to conduct a Raman microscope: 100 m slit, Ar laser (532 nm), 10 mW at the sample, a grating of 1800 grit/mm, a confocal hole of 1000 μm, and a sampling duration of 10s.

The microscope's optical system independently finds and examines three locations at every implant surface site (collar/root). 20 × 35 m regions are chosen and examined at 5 m increments with a 5 s scan time in order to map the monoclinic ZrO_2 phase [56].

4.2 Phase characterization

For morphological analysis, LV-SEM is used. And for the roughness, optical profilometry is used. In X-ray diffraction (XRD), a sample from each implant is embedded in epoxy resin and sliced into longitudinal sections using a microtome while being continually cooled by water to test for the presence of tetragonal and monoclinic ZrO_2 phases across the whole implantation. The materials are smoothed using SiC paper and polished using 3 M diamond paste, and then for 3 minutes, ultrasonic cleaning in distilled water is applied [56]. The zirconia X-ray diffraction spectra is displayed in **Figure 6**. Their X-ray diffraction patterns may identify zirconia's monoclinic and metastable tetragonal phases [57].

The threaded portions of the implants' 3D surface roughness characteristics are evaluated using optical profilometry. At 10–100× magnification, an optical profiler may be utilized to look at the gap between two subsequent implant threads. The roughness metrics of a 3D surface, including arithmetic mean deviation (Sa), root average square deviation variance (Sq), and max peak to valley altitude (Rt) are then measured using Veeco Vision software [56].

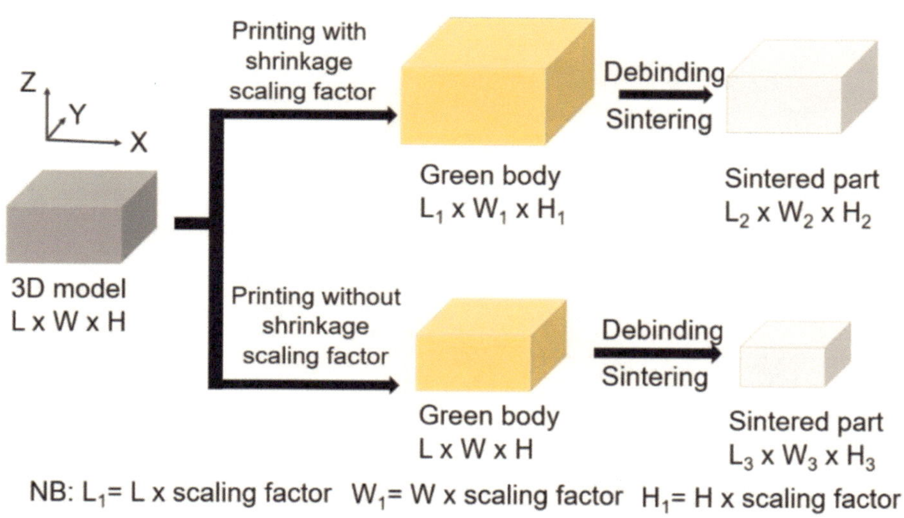

Figure 6.
Diagrammatic representation of the parameters (L, W, and H) change along the zirconia DLP pathway [44].

4.3 Mechanical testing

Using the techniques specified in **Table 2** at ambient temperature, the mechanical characteristics, including fracture toughness, Vickers hardness, and flexural strength, are thoroughly studied. A Weibull assessment of the tensile and flexural data is thus carried out to determine the possibility that zirconia would rupture. By employing the single-edge V-notched beam (SEVNB) technique and indentation, correspondingly, the fracture toughness value (K_{IC}) is assessed (**Table 3**) [44].

It is crucial to pick a reference based on the actual indentation fracture morphology since the selection of the equations also depends upon the type of crack. A femtosecond laser is used to construct an incredibly sharp V-notch (0.5 m, indicated by a red dashed line as seen in **Figure 7**) to measure fracture toughness accurately [49, 58, 59].

Figure 7 shows how U-grooves are created at particular thicknesses using a reduced-speed blade and diamond rims with a 200 m thickness under irrigation water. A 2.9 W femtosecond laser is now being used to create a crisp V-notch just at the bottom of the U-grooves. With a repeating rate of 100 kHz, the femtosecond lasers produced 290 fs linearly polarized pulses at 515 nm [50, 58].

5. Future perspectives of zirconia

5.1 Nano zirconia

Nano solids, which are tiny structures of ZrO_2, have been extensively proposed. ZrO_2 on a nanoscale, which exhibits better mechanical properties and superior biocompatibility, is frequently included into many technologies utilized in tissue engineering and dental applications [51]. Even though sophisticated powder synthesis processes can generate nano-sized particles [60–62]. Nano zirconia also showed a promising future for fuel cell applications. Cubic ZrO_2 nanoparticle production is challenging because of the range of phases that the reactions create. In a previous study, a simple precipitation method was established to adjust the form and crystallinity of cubic (Arkelite) and monoclinic (baddeleyite) zirconia nanoparticles. The

Reaction	Temperature (K)	ΔG (kJ mol^{-1})
$ZrO_2 + 2Cl_2 = ZrCl_4 + O_2$	1000	154.2
	1100	147.1
$ZrO_2 + C + 2Cl_2 = ZrCl_4 + CO_2$	1000	−241.2
	1100	−248.3
$ZrO_2 + 2C + 2Cl_2 = ZrCl_4 + 2CO$	1000	−245.7
	1100	−270.4
$ZrO_2 + 2CO + 2Cl_2 = ZrCl_4 + 2CO_2$	1000	−236.6
	1100	−226.6
$CO_2 + C = 2CO$	1000	−4.6
	1100	−22.0

Table 2.
Zirconia's carbochlorination reactions and the change in Gibbs free energy [34].

DOI: http://dx.doi.org/10.5772/intechopen.112121

	Flexural strength	Vickers hardness	Fracture toughness	
Schematic diagram	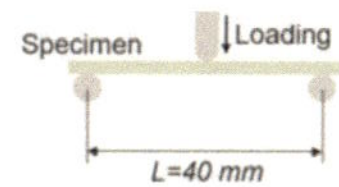	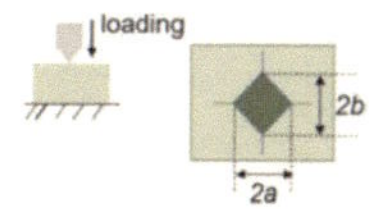	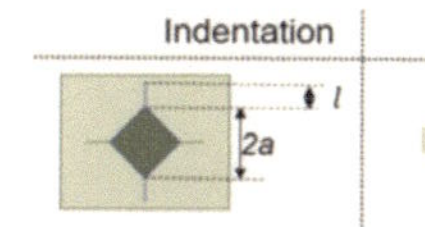	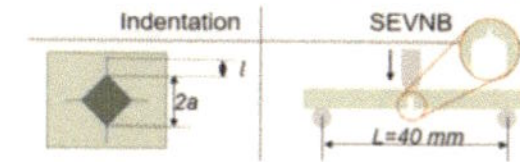
Equation	$\sigma_f = \frac{3PL}{2bh^2}$ P: Loading (N) L: Loading span (mm) b: Samples width (mm) h: Samples thickness (mm)	$HV = 0.0018544 \frac{P}{d^2}$ P: Loading (N) d: Average diagonal length of the indentation mark (mm)	$K_{IC} = 0.0089\left(\frac{E}{H}\right)^{2/5} \frac{P}{al^{1/2}}$ E: Young's modulus (210 GPa) H: Hardness (GPa) P: Indentation lead (N) a: Half of diagonal length (m) l: Crack length (m)	$K_{IC} = \frac{PL \times 10^{-6}}{BW^{3/2}} Y \frac{3\alpha^{1/2}}{2\alpha^{3/2}}$ P: Loading (N) L Loading span (mm) B: Samples thickness W: Samples width A: Notch depth/sample width $\beta = 1-\alpha$ Y: Dimensionless factor
Samples	N = 40 Size: 45(L)*4(W)*3(H) mm	N = 40 Size 10(L)*10(W)*5(H) mm	N = 40 Size: 10(L)*10(W)*5(H) mm	N = 40 Size: 45(L)*4(W)*3(H) mm
Test standard	ASTM C1161-13	ASTM C1327-15	—	ASTM C1421-99
Test condition	0.5 mm/min 1 kN load cell	Loading 98 N. dwell 10s		0.5 mm/min, 1 kN load cell
Instrument	Universal testing machine (LABSANS Shenzhen, China)	Vickers hardness tester (THV-50MDX Shanghai, China)		Universal testing machine

Table 3.
Investigation techniques for mechanical properties [44].

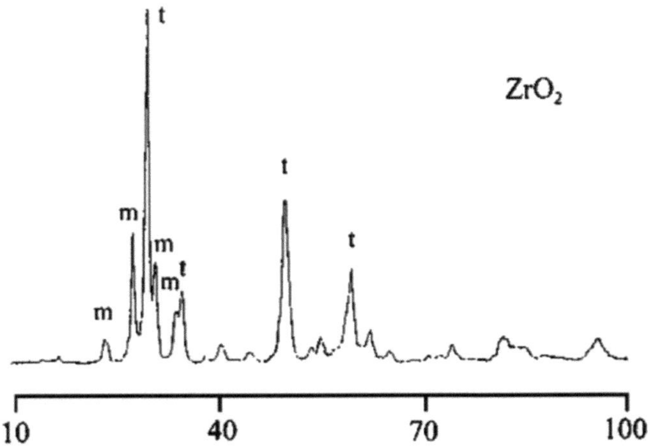

Figure 7.
XRD spectra of zirconia [57].

manufactured cubic and monoclinic zirconia can be put into the Nafion® membrane to increase the fuel cell's efficiency. The polytetrafluoroethylene backbone and perfluorinated ether sidechain are characteristics of the proton conductor electrolyte Nafion®. Additionally, due to the transpiration of water that leads the membrane to buckle, Nafion® loses conductivity at a higher temperature of 100 C [31].

Toward the latter situation, the rise in clinical defects in the dental sector has been noticed owing to veneer chipping due to pressure accumulation throughout fabrication. Additionally, the use of ceramics' nanocrystalline structure has gained the interest of engineers since they strengthen their resistance to low-temperature deterioration (LTD) [63].

Current efforts involve synthesizing nanoparticle agglomerates within 30 mm aggregates. The high-velocity oxy-fuel (HVOF) method can be subjected to spray nano-zirconia powders made in radio frequency (RF) plasma. This method employs relatively low flame temperatures (<3000°C) to ensure that the particles are only a little heated. The nano-zirconia particle-reinforced coatings are examined using scanning electron microscopy (SEM) and transmission electron microscopy (TEM). The effects of the post-spray treatment on the coatings are investigated [62].

Wet-chemical synthesis techniques, such as co-precipitation, hydrothermal synthesis, and sol–gel preparation, are primarily used to manufacture nano-ZrO_2 powders. The physical approach cannot achieve nanoscale results, and the gas-chemical process is too expensive to be repeated in a real-world setting [64].

The nanoparticles' surfaces are treated with various surfactants that have different attaching head groups and carbon chain lengths. Because of their distinct nature and properties, quantum dots (QDs) fall within the special category of nanomaterials. A QD is defined as an individual semiconducting nanocrystal that ranges in dimension from 2 to 10 nm. Several techniques, including soot vapor deposition, microwave synthesis, laser ablation or chemical oxidation of graphite, and graphite oxidation heat-induced oxidation of a molecular precursor, can be used to create QDs. In recent research, scientists devised a simple approach for producing a highly effective photocatalytic composite by adsorbing carbon quantum dots (CQDs) onto the surface of ZrO_2 NPs as demonstrated in **Figure 8**. The developed

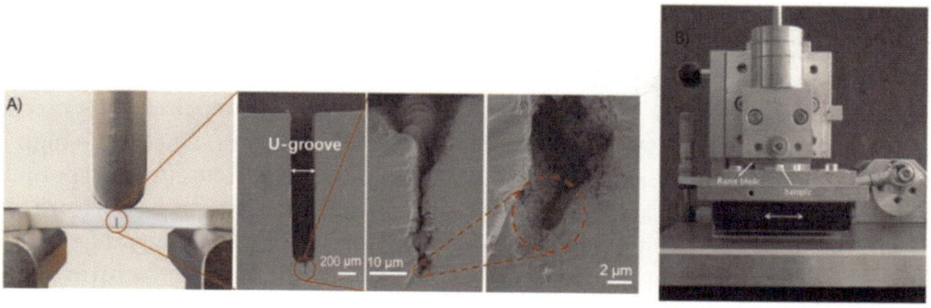

Figure 8.
(A) Diagrammatic representation of SEVNB samples. (B) Equipment for creating a razor-sharp V-notch using a blade [49, 58].

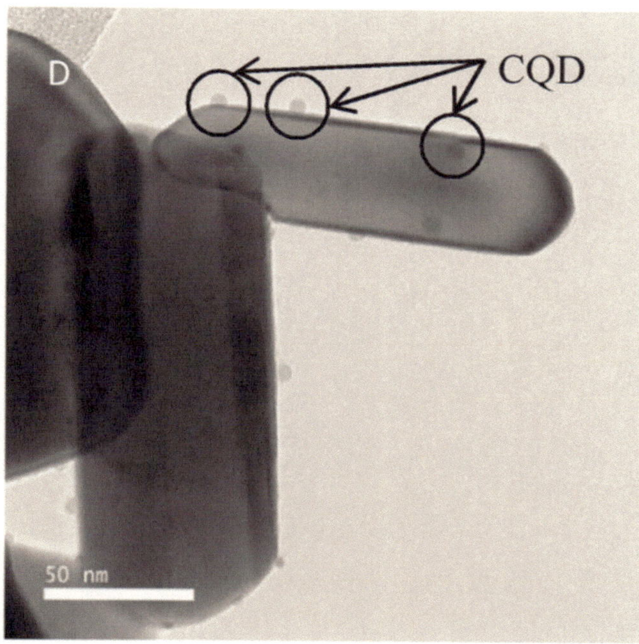

Figure 9.
TEM pictures of ZrO_2 nanoparticles with CQD embellishments [65].

photocatalyst could be useful for the rapid fading of textile colors. The composite's production method, which entailed calcination of an ammonium citrate solution, is used to make QDs. Then, using a conventional solvent-based chemical method, produced QDs are adsorbed onto the surface of ZrO_2 NPs with ultrasound assistance **Figure 9** [65].

Other studies show another development regarding a nano-structured glass-zirconia in order to increase the resilience of the interface between dental zirconia substrate and veneered porcelain. A new SiO_2–Li–Al–O_3 material to create a glass-zirconia nanostructure, prepped (SLA) glass was infiltrated through the outermost layer of fully sintered dental zirconia [66].

6. Conclusions

Zirconia has been indicated and used in various industries including, dentistry, glassware, and electronics. This comes from different reasons especially, its mechanical, electrical, thermal conductivity, and tribological properties. Due to the high demand, zirconia supply had to be produced after purifying zirconium compounds by chemical processing using acids or bases. Different processes might be performed to produce pure zirconia, such as chlorination under a high-temperature range between 623 K and 673 K, colloidal processing of zirconia, and additive manufacturing which uses a 3D printing technology with high precision and resolution. Zirconia's microstructure is analyzed by XPS and X-ray energy, phase characterization by XRD, and Mechanical properties are tested analyze the fracture toughness and flexural strength. The uses of zirconia have been constantly increasing over the last years, and the most discussed structures are nano zirconia integrated with other systems such as CQDs to enhance its characteristics.

Conflict of interest

The authors declare no conflict of interest.

Author details

Ghassan Albarghouti* and Sojood Mahmoud Farhan Darwish
Birzeit University, Ramallah, Palestine

*Address all correspondence to: gdaghra@birzeit.edu

References

[1] Yan LL, He ZY, Beier C, Klemd R. Zircon trace element constrains on the link between volcanism and plutonism in SE China. Lithos. 2018;**320–321**:28-34. DOI: 10.1016/j.lithos.2018.08.040

[2] Yakymchuk C. Prograde zircon growth in migmatites. Journal of Metamorphic Geology. 2023;**2023**

[3] Selby JH, Committee ZM, Africa S. The industrial uses of zircon and zirconia, and the radiological consequences of these. Atomic Energy. 2007;**51**:1-14

[4] Ferreira P, Ferreira A, Rocha J, Soares MR. Synthesis and structural characterization of zirconium silicates. Chemistry of Materials. 2001;**13**(2): 355-363

[5] Yugeswaran S, Selvarajan V, Dhanasekaran P, Lusvarghi L. Transferred arc plasma processing of mullite-zirconia composite from natural bauxite and zircon sand. Vacuum. 2008; **83**(2):353-359

[6] Klinger M, Kempe U, Pöpl A, Böttcher R, Trinkler M. Paramagnetic hole centres in natural zircon and zircon colouration. European Journal of Mineralogy. 2012;**24**(6):1005-1016

[7] Mineralogical Studies on Zircon [Internet]. 2009. Available from: https://www.researchgate.net/publication/277093078

[8] Aia MA. Journal of the Electrochemical Society. 1967; **114**(4):367

[9] Bona AD, Pecho OE, Alessandretti R. Zirconia as a dental biomaterial. Materials. 2015;**8**(8):4978-4991

[10] Li X, Huang J, Cao Q, Liao Y, Feng M. Investigation on polishing the concave surface of zirconia ceramics with magnetic compound fluid enhanced by hydration reaction. Magnetochemistry. 2023;**9**(3):74

[11] Tyagi B, Sidhpuria K, Shaik B, Jasra RV. Synthesis of nanocrystalline zirconia using sol-gel and precipitation techniques. Industrial and Engineering Chemistry Research. 2006;**45**(25): 8643-8650

[12] Manicone PF, Rossi Iommetti P, Raffaelli L. An overview of zirconia ceramics: Basic properties and clinical applications. Journal of Dentistry. 2007; **35**(11):819-826

[13] Piconi C, Maccauro G. Zirconia as a ceramic biomaterial. Biomaterials. 1999; **20**(1):1-25

[14] Herr C, Rizo-gorrita M, Maza-solano S, Serrera-figallo M, Torres-lagares D, Yttria-stabilized CADCAM, et al. A Review on CAD/CAM Yttria-Stabilized Tetragonal Zirconia. 2022

[15] Cadi-Essadek A, Roldan A, de Leeuw NH. Density functional theory study of the interaction of H_2O, CO_2 and CO with the ZrO_2 (111), Ni/ZrO_2 (111), YSZ (111) and Ni/YSZ (111) surfaces. Surface Science. 2016;**653**:153-162

[16] Evans AG, Marshall DB, Burlingame NH. Transformation toughening in ceramics. Advances in Ceramics. 1981;**3**(190144):202-216

[17] Basu B. Toughening of yttria-stabilised tetragonal zirconia ceramics. International Materials Reviews. 2005; **50**(4):239-256

[18] Rahmawati F, Permadani I, Syarif DG, Soepriyanto S. Electrical properties of

various composition of yttrium doped-zirconia prepared from local zircon sand. International Journal of Technology. 2017; **8**(5):939-946

[19] Saridag S, Tak O, Alniacik G. Basic properties and types of zirconia: An overview. World Journal of Stomatology. 2013;**2**(3):40

[20] De Souza RH, Kaizer MR, Borges CEP, Fernandes ABF, Correr GM, Diógenes AN, et al. Flexural strength and crystalline stability of a monolithic translucent zirconia subjected to grinding, polishing and thermal challenges. Ceramic International. 2020; **46**(16):26168-26175. DOI: 10.1016/j.ceramint.2020.07.114

[21] Hvizdoš P, Puchý V, Duszová A, Dusza J, Balázsi C. Tribological and electrical properties of ceramic matrix composites with carbon nanotubes. Ceramics International. 2012;**38**(7): 5669-5676

[22] Tanaka H, Maeda T, Narikiyo H, Morimoto T. Mechanical properties of partially stabilized zirconia for dental applications. Journal of Asian Ceramic Societies. 2019;7(4): 460-468

[23] Tong Z, Ma Q, Ni Y, Mei T, Li J, Dong G. Tribological properties of carbon fabric reinforced phenolic-based composites containing CNTs@MoS_2 hybrids. Journal of Materials Science. 2019;**54**(23):14354-14366

[24] Chen H, Zhang Y, Ding C. Tribological properties of nanostructured zirconia coatings deposited by plasma spraying. Wear. 2002;**253**(7–8):885-893

[25] Vest RW, Tallan NM. Electrical properties and zirconia: 11, Tetragonal Defect Structure of Phase and Inversion

[26] Venkataraj S, Geurts J, Weis H, Kappertz O, Njoroge WK, Jayavel R, et al. Structural and optical properties of thin lead oxide films produced by reactive direct current magnetron sputtering. Journal of Vacuum Science & Technology A: Vacuum, Surfaces, and Films. 2001;**19**(6):2870

[27] Clarke DR, Levi CG. Materials design for the next generation thermal barrier coatings. Annual Review of Materials Research. 2003;**33**:383-417

[28] Raza M. Oxygen vacancy stabilized zirconia (OVSZ); synthesis and properties What drives the phase formation of reactively sputtered transition metal oxide films ? View project. Available from: https://www.researchgate.net/publication/315142103

[29] Gautam C, Joyner J, Gautam A, Rao J, Vajtai R. Zirconia based dental ceramics: Structure, mechanical properties, biocompatibility and applications. Dalton Transactions. 2016; **45**(48):19194-19215

[30] Mistarihi Q, Umer MA, Kim JH, Hong SH, Ryu HJ. Fabrication of ZrO_2-based nanocomposites for transuranic element-burning inert matrix fuel. Nuclear Engineering and Technology. 2015;**47**(5):617-623

[31] Sigwadi R, Mokrani T, Dhlamini M. The synthesis, characterization and electrochemical study of zirconia oxide nanoparticles for fuel cell application. Physica B Condensed Materials. 2020; **2020**:581

[32] Ngian KF, Hartshorn AJ, Jenkins DH. Process for the Purification of Zirconium Compounds. Vol. 191989

[33] Movahedian A, Raygan S, Pourabdoli M. The chlorination kinetics of zirconium dioxide mixed with carbon

black. Thermochimica Acta. 2011;**512**(1–2):93-97

[34] Landsberg A, Hoatson CL, Block FE. The Chlorination Kinetics of Zirconium Dioxide in the Presence of Carbon and Carbon Monoxide

[35] Rami ML, Meireles M, Cabane B, Guizard C. Colloidal stability for concentrated zirconia aqueous suspensions. Journal of the American Ceramic Society. 2009;**2009**

[36] Aboras M, Muchtar A, Azhari CH, Yahaya N, Mah JCW. Enhancement of the microstructural and mechanical properties of dental zirconia through combined optimized colloidal processing and cold isostatic pressing. Ceramic International. 2019;**45**(2):1831-1836. DOI: 10.1016/j.ceramint.2018.10.072

[37] Lange FF. Densification of powder compacts: An unfinished story. Journal of the European Ceramic Society. 2008; **28**(7):1509-1516

[38] Ardekani KT, Ahangari H, Leila F. Marginal and internal fit of CAD/CAM and slip-cast made zirconia copings. Journal of Dental Research, Dental Clinics, Dent Prospects. 2012;**6**(2):42-48. Available from: http://dentistry.tbzmed.ac.ir/joddd

[39] Xie Z, Ma J, Xu Q, Huang Y, Cheng YB. Effects of dispersants and soluble counter-ions on aqueous dispersibility of nano-sized zirconia powder. Ceramics International. 2004; **30**(2):219-224

[40] Kaga T, Application F, Data P, Andrews PE, Melvyn J. United States Patent (19). 1991;(19)

[41] Fu X, Zou B, Xing H, Li L, Li Y, Wang X. Effect of printing strategies on forming accuracy and mechanical properties of ZrO_2 parts fabricated by SLA technology. Ceramics International. 2019;**45**(14):17630-17637

[42] Kang JH, Sakthiabirami K, Kim HA, Hosseini Toopghara SA, Jun MJ, Lim HP, et al. Effects of UV absorber on zirconia fabricated with digital light processing additive manufacturing. Materials. 2022; **15**(24)

[43] Coppola B, Schmitt J, Lacondemine T, Tardivat C, Montanaro L, Palmero P. Digital light processing stereolithography of zirconia ceramics: Slurry elaboration and orientation-reliant mechanical properties. Journal of the European Ceramic Society. 2022;**42**(6):2974-2982

[44] Sun J, Chen X, Wade-Zhu J, Binner J, Bai J. A comprehensive study of dense zirconia components fabricated by additive manufacturing. Additive Manufacturing. 2021;**2021**:43

[45] Saikumari N, Dev SM, Dev SA. Effect of calcination temperature on the properties and applications of bio extract mediated titania nano particles. Scientific Reports. 2021;**11**(1)

[46] Rodaev VV, Razlivalova SS, Zhigachev AO, Vasyukov VM, Golovin YI. Preparation of zirconia nanofibers by electrospinning and calcination with zirconium acetylacetonate as precursor. Polymers (Basel). 2019;**11**(6)

[47] Pavlik RS, Holland HJ, Payzant EA. Thermal Decomposition of Zircon Refractories

[48] Saricimen H, Ul-Hamid A, Quddus A. Effect of calcination on nanoscale zirconia produced by high temperature hydrolysis. Protection of Metals and Physical Chemistry of Surfaces. 2015; **51**(5):803-811

[49] Li R, Wang C, Ma SQ, Liu ZH, Zang CC, Zhang WY, et al. High bonding strength between zirconia and composite resin based on combined surface treatment for dental restorations. Journal of Applied Biomaterials and Functional Materials. 2020;**2020**:18

[50] Russo DS, Cinelli F, Sarti C, Giachetti L. Adhesion to zirconia: A systematic review of current conditioning methods and bonding materials. Dentistry Journal. 2019;**7**

[51] Schünemann FH, Galárraga-Vinueza ME, Magini R, Fredel M, Silva F, Souza JCM, et al. Zirconia surface modifications for implant dentistry. Materials Science and Engineering C. 2019;**98**:1294-1305

[52] Shen D, Wang H, Shi Y, Su Z, Hannig M, Fu B. The effect of surface treatments on zirconia bond strength and durability. Journal of Functional Biomaterials. 2023;**14**(2)

[53] Rojek J, Nosewicz S, Maździarz M, Kowalczyk P, Wawrzyk K, Lumelskyj D. Modeling of a sintering process at various scales. In: Procedia Engineering. Vol. 2017. 2017. pp. 263-270

[54] Grossin D, Montón A, Navarrete-Segado P, Özmen E, Urruth G, Maury F, et al. A review of additive manufacturing of ceramics by powder bed selective laser processing (sintering/melting): Calcium phosphate, silicon carbide, zirconia, alumina, and their composites. Open Ceramics. 2021;**2021**:5

[55] Patil A, DJ D, Bomze D, Gopal V. Wear behaviour of lithography ceramic manufactured dental zirconia. BMC Oral Health. 2023;**23**(1):276. Available from: https://bmcoralhealth.biomedcentral.com/articles/10.1186/s12903-023-02974-4

[56] Zinelis S, Thomas A, Syres K, Silikas N, Eliades G. Surface characterization of zirconia dental implants. Dental Materials. 2010;**26**(4):295-305

[57] Zhang QH, Feng YQ, Shi-Lu D. Preparation and Characterization of Zirconia-Silica and Zirconia-Magnesia Supports for Normal-Phase Liquid Chromatography

[58] Yang J, Huang Y, Meier LP, Wyss H, Tervoort E, Gauckler LJ. Direct coagulation casting via increasing ionic strength. In: Key Engineering Materials. Trans Tech Publications Ltd; 2002. pp. 631-636

[59] Xu J, Gan K, Yang M, Qu Y, Wu J, Yang J. Direct coagulation casting of yttria-stabilized zirconia using magnesium citrate and glycerol diacetate. Ceramics International. 2015; **41**(4):5772-5778

[60] Strecker K, Ribeiro S, Hoffmann MJ. Fracture toughness measurements of LPS-SiC: A comparison of the indentation technique and the SEVNB method. Materials Research. 2005;**8**

[61] Ćorić D, Renjo MM, Ćurković L, Žmak I. Indentation Fracture Toughness of Y-TZP Dental Ceramics

[62] Li H, Khor KA, Kumar R, Cheang P. Characterization of hydroxyapatite/nano-zirconia composite coatings deposited by high velocity oxy-fuel (HVOF) spray process. Surface and Coating Technology. 2004;**182**(2–3): 227-236

[63] Kocjan A, Pouchly V, Shen Z. Processing of zirconia nanoceramics from a coarse powder. Journal of the European Ceramic Society. 2015;**35**(4): 1285-1295

[64] Hu C, Sun J, Long C, Wu L, Zhou C, Zhang X. Synthesis of nano zirconium oxide and its application in dentistry.

Nanotechnology Reviews. 2019;**8**(1): 396-404

[65] Ratnayake SP, Mantilaka MMMGPG, Sandaruwan C, Dahanayake D, Murugan E, Kumar S, et al. Carbon quantum dots-decorated nano-zirconia: A highly efficient photocatalyst. Applied Catalysis A: General. 2019;(570):23-30

[66] Zhou M, Zhang X, Zhang Y, Li D, Zhao Z, Wang Q, et al. Construction of nanostructured glass-zirconia to improve the Interface stability of dental bilayer zirconia. Nanomaterials. 2023; **13**(4)

Section 4

Applications of Zirconia

Chapter 5

Zirconia in Restorative Dentistry

Hanumantha Murali Rao, Mamtha Kumaraswamy, Dhanu Thomas, Shivakumar Boraiah and Kuldeep Singh Rana

Abstract

Advancements in dental material science and technology have improved over the past decade tremendously. The demand for tissue-friendly esthetic materials has been fulfilled to a certain extent on account of the development of new materials. Most materials meet the requirements of esthetics, function and biocompatibility. They exhibit the properties of color stability, improved resistance to wear, dimensional stability and they are tissue-friendly. These new materials are able to provide desirable and optimum treatment outcomes on a long-term basis on account of their nature and advances in manufacturing and fabrication. Reinforced ceramic restorations are now completed within a very short time from start to finish—from digital impression to bonding of the restoration. Zirconia-reinforced ceramics share the top choice in materials along with lithium disilicate. The most recent zirconia has improved optical properties and has the potential to overcome its problem of increased opacity. These zirconia-based ceramics have successfully replaced even precious metal alloys and porcelain-fused-to-metal prostheses due to the above-mentioned qualities. This chapter throws light on zirconia and the different types used in dentistry, applications, methods of fabrication and clinically relevant properties.

Keywords: zirconia, yttria stabilized tetragonal zirconia, zirconia-based ceramics, dental porcelain, CAD-CAM

1. Introduction

Ceramics are inorganic substances produced by firing metallic and non-metallic constituents at a high temperature [1]. Ceramics have been widely used in dentistry for more than 100 years [2], and have since been advancing in terms of their physical, mechanical and optical properties [3]. As described by Kelly and Benetti [4], ceramics have been classified into three groups based on the glass content.

i. Predominantly glassy materials derived from feldspar (silica and alumina) and best reproduce the optical properties of enamel and dentin

ii. Particle-filled glasses to improve mechanical properties. For example, leucite was the first crystalline filler to be incorporated into feldspathic porcelain. Lithium disilicate and mica are other such fillers

iii. Polycrystalline ceramics: no glass is present and is much stronger than glass-based ceramics. They can either be pressed into oversized molds (compensates for shrinkage during firing) under pressure or blocks may be processed into restorations using CAD-CAM (Computer-Aided-Designing Computer-Aided-Machining).

The drawback is that there is no quantifying the amount of glass phase required for the ceramic to be included in either the predominantly glassy or the particle-filled glasses category. Also, polycrystalline ceramics do not contain glass, hence the classification lacks clarity. Another classification was outlined by Gracis et al. [5] who proposed a new grouping based on the formulation of the ceramics.

i. Glass-matrix ceramics: include feldspathic ceramics, synthetic ceramics, glass-infiltrated ceramics

ii. Polycrystalline ceramics: include alumina, stabilized zirconia, zirconia-toughened alumina, alumina-toughened zirconia

iii. Resin-matrix ceramics: include resin nano-ceramics, glass ceramic in a resin interpenetrating matrix, zirconia-silica ceramic in a resin interpenetrating matrix.

Zirconia is presently the most studied and researched dental material. The name "zirconium" originates from the Arabic term "Zarcon", which translates to "golden in colour." The dioxide form of zirconium (ZrO_2), known as zirconia, was first identified by a German chemist Martin Heinrich Klaproth in 1789 in a reaction product of heating some gems [6]. It was used along with other rare earth oxides as pigments in ceramics. Zirconia was first used as a biomaterial in 1969 for hip head replacement in orthopedics [3]. Crystalline zirconia occurs in three forms: monoclinic (M), tetragonal (T) and cubic (C). At room temperature, pure zirconia is monoclinic and remains stable in this phase up to 1170°C. Monoclinic zirconia is usually associated with surface microcracks, higher susceptibility to low temperature degradation (LTD) and lower reliability for use in dentistry [7]. Above this temperature, it transforms first into a tetragonal and then into the cubic phase. They transform from one phase into another, induced by a combination of different factors such as, temperature, humidity and stress [8].

It is crucial to know that all three phases differ in their properties. One that is useful in dentistry is tetragonal zirconia. The tetragonal form is stable between temperatures 1170° and 2370°C. The transformation from tetragonal to the monoclinic occurs upon cooling, and results in a volumetric *increase* of 4%. Pure unalloyed zirconia is unstable at room temperature and would shatter spontaneously and catastrophically on cooling due to $t \rightarrow m$. Hence, the tetragonal form must be stabilized to room temperature to overcome stress cracks and transformation volumetric changes on cooling [1, 7, 9]. For this reason, zirconia is doped with various oxides, such as yttrium oxide (Y2O3) or cerium oxide (CeO2) or other metallic oxides such as magnesium oxide (MgO) and calcium oxide (CaO). Significant molecular stability can be obtained with ZrO2 doped with Y2O3. It has superior mechanical properties than other combinations; although sintering is much more difficult, this is the main kind of zirconia considered for biomedical use (**Figure 1**) [9].

The concept of fracture toughness and tensile stress is central to the excellent mechanical properties of yttria-stabilized zirconia. Ceramics are fired at such high

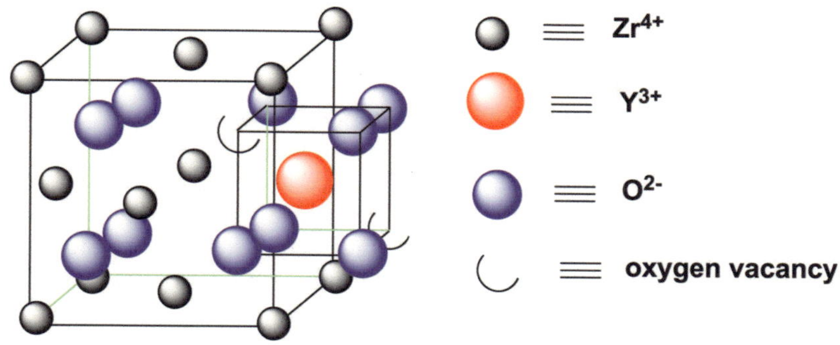

Figure 1.
Yttria-stabilized zirconia.

temperatures that upon immediate cooling, a natural crack forms, weakening the material. Fracture toughness characterizes the potential stress directed through the developed crack or flaw. When stress occurs on a zirconia surface, there is a t → m transition associated with cracking, followed by a volumetric expansion [9]. By inducing external stress or a higher tensile stress, there is an increase in volumetric expansion resulting in compressive stress around the crack tips [2] and there is an increased resistance to crack propagation. Hence, these cracks are shielded by controlled transformation from t → m and result in a gradual stabilization. The strength and fracture toughness is dependent on grain size and also the dopant concentration (yttria) [10]. This concept is called "transformation toughening", first reported by Garvie, Hannink, and Pascoe in a paper titled "Ceramic Steel?" [11]. Zirconia-based ceramics used in dentistry mostly consist of tetragonal zirconia polycrystals, partially stabilized with 3 mole % yttrium (3Y-TZP) [1].

Initially zirconia was used in root posts, orthodontic brackets dental implants. With the development of CAD/CAM, traditional metal-based crowns, prosthesis and FPD's have been replaced by improvised esthetics and enhanced tissue compatibility obtained using tooth-colored metal-free systems [3].

2. Types of zirconia in dentistry

2.1 Yttrium tetragonal zirconia polycrystals (Y-TZP)

As mentioned previously, zirconia used in dentistry usually contains 3 mol% yttria (Y2O3) as a stabilizer (3Y-TZP) is indicated for the fabrication of crowns and fixed partial dentures. The restorations are processed either by soft machining of pre-sintered blanks followed by sintering at high temperature, or by hard machining of fully sintered blocks. Yttria stabilized zirconia was first introduced in the dental market in the year 2002 via soft machining [12].

The mechanical properties of 3Y-TZP are strongly influenced by its grain size. Optimum t → m transformation occurs only in a limited range of grain size of 0.2 μm to 0.8 μm [13]. Above a critical grain size of >1 μm, 3Y-TZP is less stable and more susceptible to spontaneous t → m transformation whereas, grain sizes of <1 μm have a lower transformation rate. Below a grain size about 0.2 μm transformation is not possible, leading to reduced fracture toughness. Grain size is controlled through sintering and the sintering conditions have a strong impact on both the stability and mechanical properties of the final product. Higher sintering temperatures and longer sintering times lead to larger grain sizes [12].

Currently, available 3Y-TZP for soft machining of dental restorations employs final sintering temperatures between 1350°C and 1550°C, depending on the manufacturer. Restorations produced by soft machining are sintered after milling the restoration. This process prevents the stress-induced transformation from tetragonal to monoclinic and leads to a dense restoration and a surface virtually free of the monoclinic phase unless grinding adjustments are needed or sandblasting is performed [14].

Most manufacturers of 3Y-TZP blanks do not recommend grinding or sandblasting to avoid both the t → m transformation and the formation of surface flaws that could be detrimental to long-term performance. In contrast, restorations produced by hard machining contain a significant amount of monoclinic zirconia, usually associated with surface microcracks, higher susceptibility to LTD and lower long-term reliability [7, 12].

Other less commonly used TZPs contain 4% or 5% mol concentrations of yttria. Increasing yttria content increases the cubic phase which does not undergo transformation to the monoclinic phase. This results in a ceramic that is highly translucent but with weaker mechanical properties as transformation toughening does not occur [15, 16].

2.2 Glass-infiltrated zirconia: Toughened alumina (ZTA)

Alumina matrix is added to zirconia in order to utilize the stress-induced transformation capability of zirconia. In-Ceram Zirconia (VidentTM, Brea, CA) was introduced as a core ceramic by the addition of 33 vol% of 12 mol% ceria stabilized zirconia (12Ce-TZP) to In-Ceram Alumina [12]. It is processed by either slip casting or soft machining. In slip-casting, initial sintering takes place at 1100°C for 2 h, followed by lanthanum glass infiltration of this porous ceramic. The glass phase represents approximately 23% of the final product. The amount of porosity is greater than that of sintered 3Y-TZP and comprises between 8 and 11%, explaining the lower mechanical properties of In-Ceram® Zirconia® when compared to 3Y-TZP dental ceramics [12]. In-Ceram Zirconia is stronger and more opaque than In-Ceram Alumina and In-Ceram Spinell, therefore its use is limited to posterior crowns and fixed partial dentures [17].

2.3 Partially stabilized zirconia (PSZ)

There was a considerable amount of research on magnesia-stabilized zirconia (Mg-PSZ) for biomedical use but it was stopped in the 1990s to many causes. Mg-PSZ had higher residual porosity, which resulted in a less dense and weak final structure. The large grain sizes (30–60 μm) led to wear of the opposing structure. It also requires a higher sintering temperature of 1800°C as opposed to 1400°C of TZP, and a strictly controlled cooling cycle, especially in the aging stage. Partially stabilized zirconia usually contains a cubic matrix as the major phase and monoclinic and tetragonal zirconia as the minor phases [6]. The amount of MgO in the composition of commercial materials usually ranges between 8 and 10 mol% [18]. Precipitation of the transformable t-phase occurs during this stage, in which volume fraction is a critical factor in controlling the fracture toughness of the material [7]. Denzir-M® (Dentronic AB) is the sole Mg-PSZ ceramic available for dental restorations fabricated via hard machining [19].

Another oxide that is used to stabilize zirconia is ceria (CeO_2). Ceria provides better thermal stability and resistance to LTD than yttria-stabilized zirconia although the amount of ceria required to maintain the same stability is greater [20]. Ce^{4+} itself is unstable as such and reduces to Ce^{3+} which does not have a very good stabilizing effect on zirconia [19]. It also has a lower flexural strength but this could be improved by adding nanosized alumina in the matrix, called ceria-stabilized zirconia/alumina nanocomposite available commercially as Nanozir (Hint-Els, Griesheim, Germany) [20].

2.4 Zirconia containing lithium disilicate (ZLS)

Lithium disilicate has gained popularity as a monolith ceramic for single crowns and partial coverage restorations due to its excellent optical properties and strength and is available as machinable blanks [21]. To this technology, tetragonal zirconium oxide with a mean grit size of approximately 0.5 to 0.7 μm is added as fillers. Zirconia crystals act as a nucleating agent but remain in solution in the glassy matrix. The flexural and fracture strengths are higher than lithium disilicate glass ceramics. The mechanical properties are approximately three times higher than those determined for leucite-reinforced glass ceramics. ZLS ceramics offer an excellent combination of high strength and outstanding optical properties. Thus, these materials are interesting for the fabrication of monolithic restorations. ZLS ceramics are marketed as Vita Suprinity and Celtra Duo developed by Vita (Vita Zahnfabrik, H. Rauter GmbH & Co., Bad Säckingen, Germany) and Dentsply (Dentsply Sirona, DeguDent, GmbH, Hanau-Wolfgang, Germany), respectively [22, 23]. These materials are available as industrially prefabricated blanks for various CAD/CAM systems [7].

2.5 Resin nanoceramics

Resin-matrix ceramics also called hybrid ceramics, were specially formulated for CAD/CAM and mostly contain zirconia as fillers (5). Lava Ultimate (3 M ESPE) is a highly cross-linked polymeric matrix that contains a proprietary blend of three types of fillers: silica nanoparticles, zirconia nanoparticles and zirconia-silica nanoclusters in 80% wt. (65% vol.) [24]. This material may offer higher flexural strength and fracture toughness, which results in better long-term durability and polish over time. It is available in eight shades, in both high and low translucencies. The material is indicated for single-tooth restorations, including implant abutments [7].

Novel zirconia materials have also been chronologically divided into the following three generations:

i. The first generation, tetragonal zirconia polycrystals (3Y-TZP), consisted of 5.2 wt% or 3 mol% yttria and 0.25 wt% alumina and has a small grain size (0.3–0.5 μm), high fracture toughness (9–10 MPa/m2), high flexural strength (900–1200 MPa), and Young's modulus of 210 GPa at room temperature. It was sintered at a relatively low temperature. They exhibited high opacity due to the birefringent nature of the non-cubic phases of zirconia. However, clinical studies revealed high failure rates of fixed partial dentures, anterior and posterior crowns and zirconia implants, thus lacking mechanical integrity [14, 16].

ii. The second generation was refined drastically by eliminating the sintering aid (alumina), and increasing the sintering temperature and/or duration. The grain size of 3Y-PSZ is increased to 0.5 to 0.7 μm, and the cubic phase content is increased from 6 to 12% to 20–30%. As a consequence, the translucency is increased marginally, and the biaxial strength is decreased to 900 to 1150 MPa. Although translucency is improved, it was still insufficient for use in the anterior esthetic zone, especially single tooth restorations [16].

iii. The third generation aimed at reducing the opacity, incorporates more optically isotropic cubic zirconia (50–80%), has a grain size of 1.5 μm and is produced by increasing the yttria dopants to 4–5 mol% and increasing the sintering temperature and/or duration more than that of the second generation. This is known as 5Y-PSZ. However, cubic zirconia is weaker and more brittle than its tetragonal counterpart, which jeopardizes the strength and toughness of the zirconia. The translucency is increased significantly, but the biaxial strength is decreased to 450 to 740 MPa.

Additional experimental novel zirconia types with improved translucency have been developed, including graded zirconia and nanostructured zirconia [14, 16]. Zhang and Kim developed graded zirconia in which feldspathic glass was infiltrated into Y-TZP with improved hardness and esthetics. The graded zirconia cross-sectional structure consists of an outermost glass layer, a glass-Y-TZP layer, and a Y-TZP interior. This type of gradation occurs to have eliminated delamination of the glass [25, 26].

3. Applications of zirconia in dentistry

3.1 Dowels or posts

Generally, endodontically treated teeth have undergone significant tooth destruction, both coronal and radicular, which compromises the mechanical integrity of the tooth [27]. All endodontically treated teeth also require a build-up to replace the tooth structure that had been lost due to caries or other pathologies, and access cavity preparations [28]. An assessment is made after the treatment to determine if the prospective build-up requires additional support to be retained on the tooth in the long term. This support comes in the form of a post that is inserted from ½ to 2/3^{rds} the length of the root [29]. Posts are available in different materials such as stainless steel, nickel chromium alloys, carbon fiber and glass fiber [30]. Zirconia esthetic posts are used in the anterior teeth to overcome the black color and shadow of metal posts under translucent crowns. They have high strength and fracture toughness and are extremely hard and stiff materials which makes their removal difficult in an event that necessitates retreatment [3]. Although zirconia posts are esthetically acceptable, their clinical use is limited due to a lack of retreatability and higher stiffness than dentin which may cause functional stresses to be transferred to the dentin [31].

3.2 Crown and bridge

Missing teeth cause not only functional and structural disturbance but also influence a person's psychology and social interactions. All-ceramic restorations were limited to anterior or single-tooth restorations due to their weak mechanical strength. Porcelain-fused-to-metal has been considered as the gold standard for load-bearing restorations and multiple units till recent research focused on reinforcing ceramics which resulted in the development of lithium disilicate and oxide ceramics (alumina and zirconia) [32]. In fixed partial dentures, zirconia is used as the framework (or matrix) over which the veneering ceramic is fired. It has been found that fabricating the framework in the anatomic design rather than an arbitrary form contributes to increased strength and bonding to the veneering ceramic [33]. The cumulative survival rate for zirconia single tooth crowns was 92.7% after 3 years and for fixed partial dentures is 73.9–100% after 2–5 years [34].

3.3 Implant abutments

Dental implants have shown a high success rate for single tooth rehabilitation. Titanium, stainless steel, gold alloy, zirconia and polyether ether ketone (PEEK) are the most commonly used abutment materials in implant dentistry [35]. Titanium abutments have excellent biocompatibility and mechanical strength and are considered the gold standard for posterior regions; nonetheless, they cause a grayish discolouration of the soft tissues around the abutment [36]. In its stabilized form, zirconia ceramics show better tissue adaptation and lower plaque retention as compared to alumina and titanium when used as implant abutments. Four-year survival rate is 100%, although long-term studies show that zirconia is prone to delamination and degradation in the oral environment [3].

4. Fabrication of restorations

Zirconia restorations are fabricated using CAD/CAM by either soft machining of pre-sintered blanks or hard machining of fully sintered blanks. The conventional CAD/CAM procedure involved purely laboratory works followed by first generation developed by Duret and colleagues combining intraoral scanning and final crown produced by controlled machining and milling. As it was not widely used due to the lack of accuracy, another approach called networked CAD/CAM processing was introduced into dental technology. While computer-aided designing (or CAD) allows one to determine the 2-D geometry, computer-aided machining (or CAM) allows the processing of the proposed design to calculate the path of cutting using various tools. CAM is nothing but a machine language used by copy-milling to fabricate a ceramic prosthesis. In spite of CAD/CAM having a high success rate in terms of design and fitting, final crown fabrication and veneering are done in a dental lab [37].

Normally, there are three different approaches by which ceramic restorations are processed in dentistry: green stage processing, white stage processing and processing through hot isostatic pressing (HIP). The difference between these approaches lies in the heat treatment of the raw material used for processing. The green stage of ceramic is only ceramic powder and binder pressed into a blank. It is very soft as it is extremely porous and is not used for processing zirconia. The heat-treated (or pre-sintered) green stage gives the white stage of ceramic which is processed via soft machining. Further heat treatment of this white body results in an extremely dense (~99% theoretical density), fully sintered blank which requires custom coloring [7, 38, 39]. Sintering is a thermal treatment for bonding particles into a coherent, predominantly solid structure via mass transport events that often occur on the atomic scale. Bonding leads to improved strength and lower system energy. Sintered blanks are characterized by enhanced density and improved properties [40].

4.1 Soft machining

Soft machining of pre-sintered blanks is the most common method of processing zirconia restorations. Pre-sintered zirconia blanks are manufactured by cold-isostatic pressing (CIP) of a mixture of zirconia powder, stabilizing oxides and binding agents (the latter removed during the pre-sintering process) [21]. The prosthesis is milled from this pre-sintered block but with bigger dimensions so as to compensate for the sintering shrinkage [39]. The most common sintering method for zirconia uses conventional furnaces at temperatures between 1350°C and 1400°C and holding times ranging from 2 to 4 hours. An alternative protocol that is recommended by manufacturers using conventional ovens is a short "speed" sintering protocol which uses temperatures of 1500–1600°C and a holding time of 30 minutes that is supposed to save time and be more economical [14]. The zirconia framework attains its final mechanical properties at the end of the sintering process when it undergoes a contraction at about 25%, and reaches its correct dimensions. In order to optimize the fitting of the restoration it's imperative to know the exact volume shrinkage for every zirconia blank. The vast majority of blocks have barcodes that give information fed into the computer regarding the density of the milling block so that the framework is milled adequately oversized [39].

4.2 Hard machining

Fully sintered blocks are processed through hot isostatic pressing (HIP) at temperatures between 1400 and 1500°C. Unlike cold isostatic pressing which uses room temperature fluid under pressure to process zirconia blanks, HIP uses heat and high pressure in an argon atmosphere. This is done using special furnaces, which in addition to heavy milling procedures makes the prosthesis fabrication an expensive task [41]. Although fully dense blanks have better mechanical properties, they lack the popularity of partially sintered blanks owing to their long milling times and the hardness of the dense blanks, especially in the fabrication of fixed partial denture frameworks [38].

5. Clinical aspects

5.1 Mechanical and physical properties

Mechanical properties of the final restoration are influenced by the very first step of synthesis of the Y-TZP nanopowder. Starting powders must be crystalline, homogeneous, with high purity and narrow particle size ranges [42]. The powder is then compacted most commonly by cold isostatic pressing, to achieve a certain level of densification, followed by sintering. Many studies have been done to assess the effect of sintering temperature on the mechanical properties of zirconia and it was found that monolithic zirconia retained its biaxial flexural strength at a sintering temperature of 1550°C whereas, the biaxial flexural strength of core zirconia decreased significantly. It is established that monolithic zirconia has higher flexural strength and fracture resistance than not only conventional glass ceramics but also lithium disilicate [15]. Zirconia shows similar mechanical properties to stainless steel [6] and

Property	Flexural strength (MPa)	Fracture toughness (MPa m$^{1/2}$)	Modulus of elasticity (GPa)
3Y-TZP	900–1200	9–10	200–210
ZTA	426–688	3.1–4.61	210
Mg-PSZ	898–1118	5–10	200
ZLS	123–553	1.30–2.98	61–108.2
Resin nanoceramics	200	1.09	16.4–28.1

Table 1.
Mechanical properties of types of zirconia in dentistry.

is the highest one among ceramics used in dentistry. As the strongest and toughest of all dental ceramics, zirconia has 900–1200 MPa flexural strength, and 9–10 MPa.m$^{1/2}$ fracture toughness [43]. Some mechanical properties comparing the different types of zirconia ceramics are summarized in **Table 1**.

5.2 Optical properties of zirconia

In an all-ceramic restoration, the ceramic material may be monolithic consisting of a single ceramic material, or a ceramic core material that is covered with a ceramic veneer and is known as a bi-layered all-ceramic restoration. In the bi-layered all-ceramic restoration, the core supports the restoration and gives it strength, and the veneer provides the restoration with its final shape, shade and esthetic. However, the core may also play a part in the development of the final restoration's shade [44]. An inherent drawback of the multi-layered restoration is the chipping or delamination of the veneering ceramic, which was overcome by introducing the monolith zirconia restoration. The color of monolithic zirconia restorations is initially affected by the original shade and optical properties of zirconia ceramics determined by the manufacturing processes. Various laboratory procedures conducted to fabricate monolithic zirconia restorations may influence the color. Clinical factors such as dental background, cement and zirconia restoration features can impact the resulting color. Shade reproduction of monolithic zirconia restorations may be affected during the long process from the production of zirconia ceramic to restoration delivery. The final color of restoration is the result of manufacturing processes, laboratory procedures and clinical factors [45].

The most significant disadvantage of zirconia crowns is their relatively opaque appearance. The least translucent zirconia has 42.1% translucency of a typical glass ceramic and 72% translucency of a lithium disilicate ceramic. Two factors determine the appearance of a restoration: first, the intrinsic character and color of the material and the second is extrinsic parameters like cement layer, restoration thickness and low thermal degradation. The gloss and translucency of zirconia are brand dependent and are greatly affected by grain size and content, yttria content and the amount of impurities. Light scattering and thickness are two important factors determining the translucency of ceramic [46].

As natural zirconia is white, an immediate advantage over PFM restorations is the absence of the metallic collar usually placed on the margin, and identified by patients as a "black line" as seen with gingival recession. The high opacity of zirconia can be used as an advantage to a certain degree. The presence of discolored tooth stumps, amalgam or other heavily colored restorative materials, metallic posts and cores or carbon fiber posts may be masked up to an extent to achieve a natural appearing restoration [47]. Manufacturers have presented zirconia ceramics with different visible light transmission percentages (VLTPs) ranging from 20–50%. Five types of translucency are available for zirconia including low, medium, high, super and ultra-translucency. Low and medium translucency zirconia are commonly indicated for zirconia frameworks, whereas high, super and ultra-translucency zirconia are mostly designated for monolithic restorations [45].

Much research and development has been done to improve the optical properties of zirconia. The opacity is said to be a result of the interaction of grain sizes with the wavelength of light, refractive index mismatch between zirconia grains and the matrix, and refractive indices of the monoclinic, cubic and tetragonal phases. These factors cause light to be scattered instead of being transmitted through zirconia, leading to an opaque appearance [46]. Therefore, light scattering must be reduced to increase translucency.

5.2.1 Factors that affect zirconia translucency

i. Composition: The addition of alumina as a sintering aid to Y-TZP enhances densification creating fine-grained and homogeneous microstructures while decreasing the sintering temperature and time required [13]. Alumina also contributes to the stabilization of the tetragonal zirconia phase. However, due to the difference in the refractive index of alumina and zirconia, which at 600 nm wavelength are n = 1.76 and n = 2.21, respectively, alumina inclusions are considered significant scattering centres. Lowering the amount of alumina from 0.25 to 0.1 or 0.05 wt. % can significantly improve translucency but requires higher sintering temperatures (>1500°C) in conjunction with a longer holding time (6 h) which in turn increases the grain size, thereby leading to reduced translucency. The introduction of 0.2 mol% lanthanum oxide (La_2O_3) in conventional 0.1–0.25 wt.% Al_2O_3-doped 3Y-TZP resulted in an excellent combination of high translucency and superior hydrothermal stability while retaining excellent mechanical properties [13].

ii. Increasing the yttria dopant content to 4–5 mol% increases the cubic phase which results in an increase in the translucency, referred to as high-translucency (HT) zirconia. In 2014, 3 M ESPE presented a translucent experimental zirconia containing 7.10 wt% yttria and a reduced grain size of 150 nm (1.5 μm). This combination resulted in a highly translucent ceramic albeit the decreased transformation toughening, flexural strength and fracture toughness, all due the result of a reduced tetragonal phase (46). An even higher content of 8 mol% of yttria has been marketed as ultatranslucent zirconia [Prettau Anterior (Zirkonzahn), DD cube X2 (Dental Direkt), Katana Zirconia St and Katana Zirconia Ut Noritake (Kuraray Noritake Dental)] [46, 48].

iii. Sintering temperatures: Higher sintering temperatures report an increase in translucency, generally correlated to an associated increase in grain size, pore elimination through solid-state diffusion and consequently an increase in density. Most monolithic zirconia ceramics should be sintered in a sintering temperature between 1400 and 1550°C and no higher than that, as at temperatures of 1600 or 1700°C or after prolonged sintering, grain boundary cracks can be generated, increasing light scattering [13]. Improving color and translucency can be done by increasing the sintering temperature and time, though speed-sintered Y-TZP can acquire less desirable wear properties than normal sintered zirconia [48].

iv. Porosities: Pores of a similar size to that of visible light wavelength (400 nm to 700 nm) is the main cause of light scattering. Pores larger than 50 nm can cause significant scattering negatively affecting light transmittance. In order to minimize pore size, the starting zirconia powder needs to be on nanometric scale to give a dense zirconia blank [13]. For high-density zirconia, employing 40 nm-sized particles instead of 90 nm amplifies the sintering density, and reduces pores and scattering [13, 48].

v. Defects such as oxygen spaces: Sintering conditions have a significant effect on oxygen vacancies. When the Y-TZP is under controlled firing or sintering is done in a reduced environment, oxygen vacancies occur. This means that some of the oxygen atoms leave their respective positions in the crystal lattice creating a defect. The amount of these oxygen vacancies affect light scattering as they serve as scattering centres. On the other hand, post-sintering in an oxidizing environment could put some oxygen back into Y-TZP crystals. However, this process also creates porosity due to the combination of vacancies to produce larger ones at high temperatures. Therefore, the main consideration is to control the heat treatment to decrease the number of oxygen vacancies in the material [48, 49].

Grain size: The large grain size of tetragonal zirconia polycrystals has been correlated to reduced translucency and increased light scattering. According to classical physics, materials containing grain sizes of less than 1 μm appear less opaque due to the reflection and absorption of visible light. On the other hand, particles larger than 10 μm scatter more light and appear more opaque. Due to the birefringence (double refraction) and polycrystalline nature of zirconia, there is more scattering than transmission. Reducing the grain size of TZP zirconia improves translucency. The higher sintering temperature is associated with increased grain size but reduced strength of the ceramic. A grain size of less than 100 nm is necessary for acceptable translucency. Currently, the mean grain size of contemporary zirconia ceramics lies between 0.2 and 0.8 μm [13, 48]. To achieve a translucency comparable to dental porcelains, the mean grain size of 3Y-TZP should be about 82 nm for 1.3 mm restorative thickness, 77 nm for 1.5 mm and 70 nm for 2 mm [25].

Restoration thickness: Translucency decreases with increased restoration thickness and seems to be brand-dependent [45]. A restoration as thick as 0.5 mm has better tooth-like translucency. A minor increase or decrease in the restoration thickness can alter the translucency significantly. A minimum thickness of 0.9 mm is required for acceptable shade matching of HT zirconia. For color masking, a minimum thickness of 1 mm is required or 1.6 mm thickness for ideal masking [45, 46].

5.3 Low temperature degradation (LTD)

LTD is also called aging and is defined as a spontaneous t → m transformation occurring over time at low temperatures, and the transformation is not triggered by local stresses or an advancing crack [50]. This occurs in the presence of water and starts in isolated grains on the zirconia surface leading to an increase in volume [51]. Due to this, there are stresses created in the neighboring grains and resulting in microcrack formation which allows the water to penetrate and ultimately there is a significant decrease in the strength [20]. This transformation may be accelerated by water humidity [52], and slowed using smaller grains and higher amounts of stabilizing oxides [39]. CeO_2 in 12 and 14 mol% is more resistant to LTD than Y-TZP, however, it appears more yellow. When Ce^{4+} is reduced to Ce^{3+}, the Ce-TZP becomes dark gray due to the high concentration of oxygen vacancies. LTD in Ce-TZP may be accelerated on intake of reducing foods like glucose and lactose [50]. Other factors that influence LTD are tensile stresses, grain size and residual stressed post sintering [20]. Tensile stresses as low as 400 MPa can induce LTD over a course of 5 years. A crack tip forms on the surface of zirconia which are exposed to water (or any other fluids) and the grain transforms to monoclinic. As t → m is associated with a volume expansion, there is an uplift of the surface at the site of transformation. Under this grain, there is a large compression that subjects the underlying grains to small tensile stresses. These grains transform to monoclinic and as a cycle, continue to crack through the depth of the zirconia, thus weakening the entire structure. Reducing the grain size is said to have a restraining effect on LTD. Critical grain size for pure zirconia at room temperature falls in the range of 5–10 nm, although this number was calculated at above 100°C in various experiments [50].

5.4 Other considerations

Technical problems associated with the clinical performance of zirconia crowns and fixed dental prostheses have been reported, in particular, chipping of the veneering porcelain when applied to zirconia framework structures and loss of retention. Attempts to minimize the chipping of veneering porcelain by milling the veneers and frameworks separately and subsequently luting them with either a luting agent or using fusing firing (CAD on) have not been quite sufficient to address the chipping concerns. Another attempt to overcome the veneer chipping problem was the introduction of zirconia in the form of fully anatomical contoured monolithic prostheses intended to be used without veneering porcelain [14]. The problems of layering in zirconia-based restorations, such as veneering ceramic delamination/chipping and veneering ceramic zirconia ceramic incompatibility, do not exist in monolithic zirconia restorations [45]. Although monolithic zirconia ceramics have lower flexural strength than the framework zirconia of zirconia-based restorations of equal thicknesses, monolithic zirconia crowns have a higher fracture resistance than zirconia-based crowns due to increased zirconia thicknesses and lack of veneering ceramics [45].

Ideally, chair-side grinding and adjustment of anatomically contoured zirconia restorations should be avoided as they can produce rough surfaces. However, for many restorations, it is necessary to optimize occlusion, proximal contacts and axial contour. Grinding has two counteracting effects on zirconia: either it produces surface compressive stress that can positively enhance crack healing (by transformation toughening) and increase the material strength by transformation toughening or it can induce surface flaws that may exceed the depth of the compressive layer and negatively influence the strength of the material. Microcracks can be triggered by a number of stimuli such as thermal changes, humidity, airborne-particle abrasion and grinding, as mentioned previously [53]. There are a number of finishing and polishing systems available for zirconia restorations in different grades of diamond grits.

Aging of anatomically contoured Y-TZP restorations could be more crucial because restorations are in direct contact with oral fluid. Hydrothermal aging of zirconia, known as low-temperature degradation, can occur over time within the temperature range of 65–500°C in the presence of water and other solvents. Although this mechanism is very slow in oral temperatures, zirconia restorations are exposed to other factors such as constant humidity, thermal changes, pH fluctuation, and repeated high occlusal loads due to mastication and parafunctional habits that can accelerate the aging process and reduce the material's fracture resistance. Also, the chemical composition, the microstructure of various brands of high-translucency zirconia, the thickness of the restoration, and the processing can influence resistance to aging [53].

There are few long-term clinical trials to assess the longevity of zirconia restorations. Most studies concluded that the most common cause of failure was the chipping of the veneering ceramic. Moreover, secondary caries and marginal gaps were found in 56 to 59% of fixed partial denture abutments, although periodontal health was maintained as zirconia is shown to be highly biocompatible with gingival tissues. Zirconia implants were well-tolerated with healthy peri-implant bone and no bleeding on probing in a 2-year clinical study [54].

6. Conclusion

The innovation of newer materials in the field of dentistry has been expanding over the past decade. Most research is focused on eliminating the inherent opaque nature and susceptibility to LTD. With single-day, chairside fabrication of prosthesis through CAD/CAM, zirconia is still in its early stages and has a long way ahead to meet various criteria that make it perfectly fit for use as long-term restorations.

Author details

Hanumantha Murali Rao[1*], Mamtha Kumaraswamy[2], Dhanu Thomas[3], Shivakumar Boraiah[4] and Kuldeep Singh Rana[5]

1 Department of Conservative Dentistry and Endodontics, D.A.P.M.R.V. Dental College and Hospital, Bengaluru, Karnataka, India

2 Private Practitioner, Bengaluru, Karnataka, India

3 Conservative Dentistry and Endodontics, D A Pandu Memorial R V Dental College, Bengaluru, India

4 Government Dental College and Research Institute, Bengaluru, India

5 Conservative Dentistry and Endodontics, College of Dental Surgery, Indore, MP, India

*Address all correspondence to: drmuralihrao@yahoo.com

References

[1] Sakaguchi RL, Ferracane JL, Powers JM, editors. Craig's Restorative Dental Materials. Fourteenth ed. St. Louis, Missouri: Elsevier; 2019. p. 340

[2] Miyazaki T, Nakamura T, Matsumura H, Ban S, Kobayashi T. Current status of zirconia restoration. Journal of Prosthodontic Research. 2013;**57**(4):236-261

[3] Madfa AA, Al-Sanabani FA, Al-Qudami NH, Al-Sanabani JS, Amran AG. Use of zirconia in dentistry: An overview. Open Biomaterials Journal. 2014;**5**(1):1-7

[4] Kelly J, Benetti P. Ceramic materials in dentistry: Historical evolution and current practice: Ceramic materials in dentistry. Australian Dental Journal. 2011;**56**:84-96

[5] Gracis S, Thompson V, Ferencz J, Silva N, Bonfante E. A new classification system for all-ceramic and ceramic-like restorative materials. The International Journal of Prosthodontics. 2016;**28**(3):227-235

[6] Piconi C, Maccauro G. Zirconia as a ceramic biomaterial. Biomaterials. 1999;**20**(1):1-25

[7] Abd El-Ghany OS, Sherief AH. Zirconia based ceramics, some clinical and biological aspects: Review. Future Dental Journal. 2016;**2**(2):55-64

[8] Arellano Moncayo AM, Peñate L, Arregui M, Giner-Tarrida L, Cedeño R. State of the art of different zirconia materials and their indications according to evidence-based clinical performance: A narrative review. Dental Journal. 2023;**11**(1):18

[9] Manicone PF, Rossi Iommetti P, Raffaelli L. An overview of zirconia ceramics: Basic properties and clinical applications. Journal of Dentistry. 2007;**35**(11):819-826

[10] Garvie RC, Nicholson PS. Structure and thermomechanical properties of partially stabilized zirconia in the CaO-ZrO2 system. Journal of the American Ceramic Society. 1972;**55**(3):152-157

[11] Hannink RHJ, Kelly PM, Muddle BC. Transformation toughening in zirconia-containing ceramics. Journal of the American Ceramic Society. 2004;**83**(3):461-487

[12] Denry I, Holloway J. Ceramics for dental applications: A review. Materials. 2010;**3**(1):351-368

[13] Kontonasaki E, Rigos AE, Ilia C, Istantsos T. Monolithic zirconia: An update to current knowledge. Optical properties, Wear, and clinical performance. Dental Journal. 2019;**7**(3):90

[14] Ahmed WM, Troczynski T, McCullagh AP, Wyatt CCL, Carvalho RM. The influence of altering sintering protocols on the optical and mechanical properties of zirconia: A review. Journal of Esthetic and Restorative Dentistry. 2019;**31**(5):423-430

[15] Alqutaibi AY, Ghulam O, Krsoum M, Binmahmoud S, Taher H, Elmalky W, et al. Revolution of current dental zirconia: A comprehensive review. Molecules. 2022;**27**(5):1699

[16] Zhang Y, Lawn BR. Novel zirconia materials in dentistry. Journal of Dental Research. 2018;**97**(2):140-147

[17] Shen C, Rawls HR, Esquivel-Upshaw JF. Phillips' Science of Dental Materials. 13th ed. Philadelphia: Elsevier, Inc; 2021

[18] Sundh A, Sjogren G. Fracture resistance of all-ceramic zirconia bridges with differing phase stabilizers and quality of sintering. Dental Materials. 2006;**22**(8):778-784

[19] Bajraktarova-Valjakova E, Korunoska-Stevkovska V, Kapusevska B, Gigovski N, Bajraktarova-Misevska C, Grozdanov A. Contemporary dental ceramic materials, a review: Chemical composition, physical and mechanical properties, indications for use. Open Access Macedonian Journal of Medical Sciences. 2018;**6**(9):1742-1755

[20] Li RWK, Chow TW, Matinlinna JP. Ceramic dental biomaterials and CAD/CAM technology: State of the art. Journal of Prosthodontic Research. 2014;**58**(4):208-216

[21] Zarone F, Di Mauro MI, Ausiello P, Ruggiero G, Sorrentino R. Current status on lithium disilicate and zirconia: A narrative review. BMC Oral Health. 2019;**19**(1):134

[22] Zarone F, Ruggiero G, Leone R, Breschi L, Leuci S, Sorrentino R. Zirconia-reinforced lithium silicate (ZLS) mechanical and biological properties: A literature review. Journal of Dentistry. 2021;**109**:103661

[23] Elsaka SE, Elnaghy AM. Mechanical properties of zirconia reinforced lithium silicate glass-ceramic. Dental Materials. 2016;**32**(7):908-914

[24] Chen C, Trindade FZ, de Jager N, Kleverlaan CJ, Feilzer AJ. The fracture resistance of a CAD/CAM resin Nano ceramic (RNC) and a CAD ceramic at different thicknesses. Dental Materials. 2014;**30**(9):954-962

[25] Zhang Y. Making yttria-stabilized tetragonal zirconia translucent. Dental Materials. 2014;**30**(10):1195-1203

[26] Zhang Y, Kim JW. Graded structures for damage resistant and aesthetic all-ceramic restorations. Dental Materials. 2009;**25**(6):781-790

[27] Kishen A. Mechanisms and risk factors for fracture predilection in endodontically treated teeth. Endodontic Topics. 2006;**13**(1):57-83

[28] McLean A. Criteria for the predictably restorable endodontically treated tooth. Journal-Canadian Dental Association. 1 Oct 1998;**64**:652-656

[29] Stockton LW. Factors affecting retention of post systems: A literature review. The Journal of Prosthetic Dentistry. 1999;**81**(4):380-385

[30] Schwartz R, Robbins J. Post placement and restoration of Endodontically treated teeth: A literature review. Journal of Endodontia. 2004;**30**(5):289-301

[31] Özkurt Z, Iseri U, Kazazoglu E. Zirconia ceramic post systems: A literature review and a case report. Dental Materials Journal. 2010;**29**(3):233-245

[32] Saeed F, Muhammad N, Khan AS, Sharif F, Rahim A, Ahmad P, et al. Prosthodontics dental materials: From conventional to unconventional. Materials Science and Engineering: C. 2020;**106**:110167

[33] Raigrodski AJ, Hillstead MB, Meng GK, Chung KH. Survival and complications of zirconia-based fixed

dental prostheses: A systematic review. The Journal of Prosthetic Dentistry. 2012;**107**(3):170-177

[34] Triwatana P, Nagaviroj N, Tulapornchai C. Clinical performance and failures of zirconia-based fixed partial dentures: A review literature. Journal of Advanced Prosthodontics. 2012;**4**(2):76

[35] Shafie HR, White BA. Implant abutment materials. In: Shafie HR, editor. Clinical and Laboratory Manual of Dental Implant Abutments [Internet]. Oxford, UK: John Wiley & Sons, Ltd; 2014. pp. 1-16 [cited 2023 Mar 16] Available from: https://onlinelibrary.wiley.com/doi/10.1002/9781118928547.ch1

[36] Totou D, Naka O, Mehta SB, Banerji S. Esthetic, mechanical, and biological outcomes of various implant abutments for single-tooth replacement in the anterior region: A systematic review of the literature. International Journal of Implant Dentistry. 2021;**7**(1):85

[37] Spitznagel FA, Boldt J, Gierthmuehlen PC. CAD/CAM ceramic restorative materials for natural teeth. Journal of Dental Research. 2018;**97**(10):1082-1091

[38] Beuer F, Schweiger J, Edelhoff D. Digital dentistry: An overview of recent developments for CAD/CAM generated restorations. British Dental Journal. 2008;**204**(9):505-511

[39] Nistor L, Grădinaru M. Zirconia use in dentistry-manufacturing and properties. Current Health Sciences Journal. 2019;**1**:28-35

[40] Helvey GA. Classification of dental ceramics. Inside Continuing Education. Apr 2013;**13**:62-68

[41] Grech J, Antunes E. Zirconia in dental prosthetics: A literature review. Journal of Materials Research and Technology. 2019;**8**(5):4956-4964

[42] Kontonasaki E, Giasimakopoulos P, Rigos AE. Strength and aging resistance of monolithic zirconia: An update to current knowledge. Japanese Dental Science Review. 2020;**56**(1):1-23

[43] Malkondu Ö, Tinastepe N, Akan E, Kazazoğlu E. An overview of monolithic zirconia in dentistry. Biotechnology and Biotechnological Equipment. 2016;**30**(4):644-652

[44] Warreth A, Elkareimi Y. All-ceramic restorations: A review of the literature. Saudi Dental Journal. 2020;**32**(8):365-372

[45] Tabatabaian F. Color aspect of monolithic zirconia restorations: A review of the literature: Color in monolithic zirconia restorations. Journal of Prosthodontics. 2019;**28**(3):276-287

[46] Ghodsi S. A review on translucent zirconia. The European Journal of Prosthodontics and Restorative Dentistry. 2018;**26**:62-74

[47] Vichi A, Louca C, Corciolani G, Ferrari M. Color related to ceramic and zirconia restorations: a review. Dental materials. 1 Jan 2011;**27**(1):97-108

[48] Pekkan G, Pekkan K, Bayindir BÇ, Özcan M, Karasu B. Factors affecting the translucency of monolithic zirconia ceramics: A review from materials science perspective. Dental Materials Journal. 2020;**39**(1):1-8

[49] Johnson WW. Fundamentals of operative dentistry: A contemporary approach. Journal of Prosthodontics. 2002;**11**(3):224-225

[50] Lughi V, Sergo V. Low temperature degradation -aging- of zirconia: A critical review of the relevant aspects in dentistry. Dental Materials. 2010;**26**(8):807-820

[51] Chevalier J. What future for zirconia as a biomaterial? Biomaterials. 2006;**27**(4):535-543

[52] Chevalier J, Gremillard L, Virkar AV, Clarke DR. The tetragonal-monoclinic transformation in zirconia: Lessons learned and future trends. Journal of the American Ceramic Society. 2009;**92**(9):1901-1920

[53] Khayat W, Chebib N, Finkelman M, Khayat S, Ali A. Effect of grinding and polishing on roughness and strength of zirconia. The Journal of Prosthetic Dentistry. 2018;**119**(4):626-631

[54] Ãzkurt Z, KazazoÄlu E. Clinical success of zirconia in dental applications. Journal of Prosthodontics. 2010;**19**(1):64-68

Chapter 6

Recent Modifications of Zirconia in Dentistry

Ghassan Albarghouti and Haneen Sadi

Abstract

In restorative dentistry, there are basically two requirements aspired to be fulfilled by the material of choice to be the main constituent of the restorations, those include superior mechanical characteristics and outstanding esthetic properties. Zirconia (ZrO_2) attains great popularity nowadays and is considered a promising material in dental applications. The excellent tensile strength, high thermal stability, relatively low thermal conductivity, wear resistance, corrosion resistance, chemical stability, low cytotoxicity, minimal bacterial adhesion, and biocompatibility properties of zirconia adding to them its tooth-like color and esthetic appearance have promoted its introduction as a successive dental substance. It was found to be a potential alternative and favorable material in dental restorations competing with many of the previously known and employed ceramics and metals, such as titanium. Despite the excellent properties and wide use of titanium in dental applications, it still suffers from unfavorable drawbacks. However, some problems in zirconia diminish its mechanical properties, such as phase transformation and aging, which could be overcome *via* the utilization of dopants within the zirconia's structure. This chapter discussed the main stabilized zirconia types, properties, dental components, manufacturing, and treatment techniques. Further modifications on zirconia with the maintenance of both mechanical and esthetic properties are still under investigation.

Keywords: zirconia, ceramics, dental applications, properties, surface modifications

1. Introduction

Zirconium, a strong solid group (IV) transition metal, was isolated for the first time as an impure metal in 1824 by Berzelius and produced as a pure metal by van Arkel and de Boer in 1925 [1]. It has a hexagonal close-packed (HCP) crystalline structure at ambient conditions [2]. Zirconium is well-known for its ductility, malleability, and ease of forming stable compounds. Therefore, it occurs in nature in conjunction with other components as ores, not as a pure metal. The most common zirconium ores include the more abundant but less pure zirconium(IV) silicate ($ZrSiO_4$), commonly known as zircon, and the relatively pure baddeleyite which contains 96.5 to 98.5 weight percent zirconium dioxide also named zirconia [3]. Some of the main chemical and physical characteristics of zirconium are presented in **Table 1**. The importance of zirconium manifests clearly through its compounds,

Property	Value
Atomic Symbol	Zr
Atomic Number	40
Atomic mass	91.224
Density (g/cm^3)	6.49
Melting Point (°K)	2128
Boiling Point (°K)	4682

Table 1.
Some physical properties of zirconium.

which have several interesting applications. Some of these applications are associated with the nuclear field in the form of zircaloys. Moreover, zirconium in organometallic compounds, such as zirconocenes, could act as an intermediate in the synthesis of several biological organic compounds; and zirconia ceramics, which will be discussed in depth in this chapter [4, 5].

Zirconia (ZrO_2) is a white polycrystalline ceramic material; it is one of the oxide ceramics with superior mechanical and biomedical properties. The crystallographic form of zirconia is pressure and temperature-dependent, which means it can assume one of three phases: monoclinic (m), tetragonal (t), or cubic form (c). At ambient pressure, the monoclinic structure, with the shape of a deformed rectangular prism, is the stable form at room temperature, however, it has inferior mechanical performance when compared with other phases. Crystals assume the monoclinic structure until 1170°C, where the unusual performance occurs: shrinkage upon heating, resulting in a transition to the metastable tetragonal shape, a regular rectangular prism with improved mechanical properties. This structure is stable between 1170 and 2370°C; above this temperature, further shrinkage occurs to form the stable cubic structure, a form with moderate mechanical properties [6]. Such phase transitions are summarized in **Figure 1**.

The unusual performance of temperature-dependent crystalline structural transitions presents a considerable problem upon cooling. A phase transformation from the metastable tetragonal into the monoclinic phase ($t \rightarrow m$) occurs upon cooling with volume expansion of about (3–5%), the lattice becomes rigid and unable to accommodate the rapid volume expansion. Such considerable stresses through the lattice result in a hysteresis behavior of zirconia, where catastrophic fracture and propagation of any lattice cracks ensue over time. Therefore, pure zirconia is not easily produced and employed due to spontaneous structural failures [7, 8]. In the case of zirconia-based biomaterials, this problem could be catastrophic.

For the sake of inhibiting ($t \rightarrow m$) transformation or at least reducing the transformation rate and maintaining the metastable tetragonal lattice at room temperature, which in turn inhibits crack propagation, some significant factors could be controlled [7, 8]. The addition of stabilizers in sufficient concentrations allows the tetragonal lattice to maintain at ambient conditions.

Zirconia stabilization is a process that aims to suppress its crystalline structure during cooling; by adding specific amounts of dopants to zirconia to avoid the conversion to the monoclinic form during cooling and stabilize the tetragonal form at room temperature to a great degree. Dopants, also called stabilizers, are metallic oxides such as magnesium, calcium, cerium, and yttrium oxides.

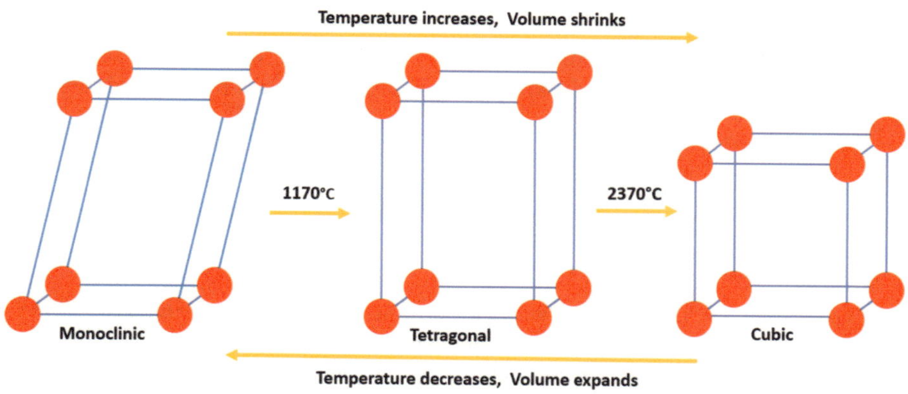

Figure 1.
Phase transformations of zirconia crystalline structure as temperature changes.

The lattice after stabilization overcomes crack propagation through a mechanism known as "transformation toughening" in which the toughness is enhanced because of the dissipation of the energy when cracking initiates to overcome the effect of the stabilizer, a (t → m) transition occurs followed by volume expansion, which leads to close the crack and blunt any developing crack [6]. This phenomenon would increase the flexural strength of the lattice.

Stabilized zirconia takes a unique place among oxide ceramics due to its distinctive mechanical and biological properties. Its mechanical properties are comparable to those of stainless steel, for example, excellent tensile strength, high thermal stability, relatively low thermal conductivity, wear resistance, and corrosion resistance. Nevertheless, the excellent chemical stability, low cytotoxicity, minimal bacterial adhesion, biocompatibility, and tooth-like color "ivory" of zirconia make it a material of keen interest in several biomedical applications, substantially in dentistry.

Zirconia was used in orthopedic joint replacement, such as hip joint replacement and surgical implants. In dentistry, with its protruding mechanical properties, zirconia is successfully used in endosseous implants with the maintenance of the natural ivory look of the tooth. Its esthetic properties enhance its utilization in dental applications widely; since esthetic concerns become a priority, as well as medication aspects as in **Figure 2** [8].

Zirconia aging or low-temperature degradation (LTD) negatively affects its properties, it is a water-catalyzed phenomenon that results in a slow (t→m) transformation and does not require mechanical stress. The mechanism of zirconia aging is similar to that of transformation toughening where the transformation is prolonged

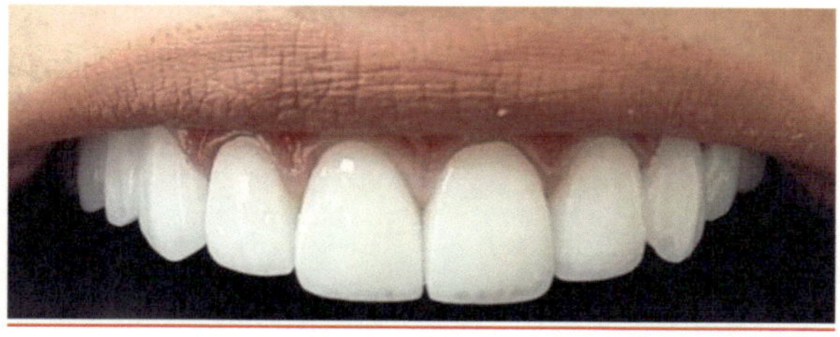

Figure 2.
Zirconia's esthetic properties in dentistry.

with volume expansion and stress induction to the surrounding. Then, the surface could rise, allowing for water to move through the lattice and penetrate down, which would exacerbate surface degradation [9]. This problem could be triggered by surface roughness.

2. Types of stabilized zirconia ceramics available for dental and biomedical purposes

As a ceramic material, zirconia was found to be hard but brittle with inferior impact resistance; this constrained its application and dictated the modification of such material with promising characteristics to permit it to work as an alternative for several metal-based artificial joints and implants. Moreover, materials to be employed in dental aspects have to be durable and stable under harsh oral cavity conditions [10]. The stabilization processes of zirconia ceramics, in which oxides are added to zirconia, enhanced its phases' stability and retained the preferred tetragonal structure more than before at ambient temperature, resulting in partially stabilized zirconia (PSZ), zirconia toughened ceramics, or tetragonal zirconia polycrystals (TZP) with improved properties [11]. Even though there are many types of stabilized zirconia combinations, only three of them, yttrium tetragonal zirconia polycrystals (Y-TZP), magnesium partially stabilized zirconia (Mg-PSZ), and zirconia toughened alumina (ZTA), are intensively tested and introduced in dentistry to overcome ($t \rightarrow m$) transformation. The mechanical properties of these types are summarized in **Table 2**.

2.1 Yttrium tetragonal zirconia polycrystals (3Y-TZP)

When Yttria (Y_2O_3) is the dopant, a monophasic tetragonal zirconia polycrystal (TZP) is obtained, usually containing 3 mol% of yttrium oxide (3Y-TZP). This type of low porosity and relatively high-density stabilized zirconia attracted considerable attention when compared with other combinations due to its chemical stability and better mechanical properties supporting its biocompatibility. Exhibiting the transformation toughening phenomenon makes it a favorable type in medical applications and particularly dental implementations, including implants, crowns, abutments, bridges, and fixed partial dentures. Production of dental restorations is carried out through hard or soft machining along with sintering processes (Processing techniques will be discussed in a separate section of this chapter). The key point of tetragonal phase stability at ambient temperature is the vacancies in the zirconia lattice appeared due to the presence of yttria.

	3Y-TZP	ZTA	Mg-PSZ
Chemical components	Y_2O_3, ZrO_2	Al_2O_3, ZrO_2	MgO, ZrO_2
Crystalline phase	Monophase	Biphase	Biphase
Flexural strength (MPa)	800–1300	750–850	700–800
Hardness (GPa, Vicker's)	10–12	12–15	5–16
Fracture toughness (MPam $^{1/2}$)	5–10	6–12	8–15

Table 2.
The main mechanical properties of the principal types of stabilized zirconia [8].

Despite the features, (3Y-TZP) undergoes aging or LTD phenomenon under hydrothermal conditions at low temperatures as well as human body temperature, and as a result, its mechanical properties degrade. In fact, the vacancies in the stabilized zirconia could act as host sites for water molecules diffused from the surrounding under hydrothermal conditions, decreasing site numbers consequently, as well as the stability of the tetragonal phase. This phenomenon is induced by exposure to hydrothermal "intraoral" conditions, surface roughness, microcracking, and stresses. It is found to be sensitive to processing conditions, including mixing, distribution, and polishing techniques. Some factors can be controlled to impact the mechanical properties of stabilized zirconia, including the grain size and sintering conditions. Spontaneous (t → m) transformation is susceptible unless the grain size is small. The transformation rate is reduced when the grain size is less than 1 μm and prevented when it is less than about 0.2 μm, and thus the brittleness of zirconia will reduce. The grain size is directly affected by sintering conditions, particularly sintering time and temperature. Large grain size will be obtained after a long time or high sintering temperature affecting the phase stability [9]. To overcome the LTD problem and failure of zirconia and improve its mechanical properties, efforts focused on the production of new composites, alumina is introduced instead of yttria to have zirconia toughened alumina. But improved (3Y-TZP) is still an attractive composite especially when computer-aided manufacturing and computer-aided design (CAM/CAD) processing are applied. Some generations reduced the LTD by adding a small amount of alumina to the (3Y-TZP) composite [12].

2.2 Zirconia-toughened alumina (ZTA)

A biphasic ZTA is accomplished by combining zirconia with an alumina matrix by slip casting or soft machining processes for the sake of integration between alumina's high stiffness with zirconia's superior toughness [13]. This type of stabilized zirconia was found to have better fracture toughness and wear resistance, also when compared with (3Y-TZP) it has enhanced aging resistance [14]. There is no stabilizing dopant in this type. Thus, the stability of the tetragonal phase at ordinary temperatures depends mainly on the particles' morphology and size, in addition to whether the location is intergranular or intragranular particles [9]. By adding zirconia, the grain growth of alumina is inhibited, improving the fracture toughness of the composite. An advantage of this combination is that both of these components are naturally white and thus act as an efficient mask for teeth of dark colors. However, ZTA exhibits low translucency. Therefore, it is not ideal in cases where esthetic concerns are paramount. The opalescence of ZTA precluded it from dental restorations of anterior sites [12, 15].

For better outcomes, ZTA was incorporated into new composites using several additives. Many ceramic oxides are utilized within ZTA to modify their properties by influencing the lattice parameters. For example, a newly developed bio-safe ternary composite of zirconia, alumina, and titania (TiO_2) facilitates sintering while processing and attains high mechanical properties with low manufacturing costs [13]. In several studies, alumina was used as an additive with other stabilized zirconia types. Alumina nanoparticles enhanced flexural strength and fracture toughness when incorporated with ceria-stabilized TZP [10]. The addition of a specific amount of silver as a dopant to ZTA enhances the antibacterial properties of the dental parts [16].

The two possible zirconia-alumina composites: ZTA, which is obtained when the alumina matrix is reinforced with ZrO_2 particles (alumina is the main component in ZTA), and alumina-toughened zirconia (ATZ), which is obtained when the zirconia matrix is reinforced with Al_2O_3 particles. Both composites show better toughness; ATZ displays improved mechanical stability and aging resistance, while ZTA exhibits much better-aging resistance [17–20].

2.3 Magnesium partially stabilized zirconia (Mg-PSZ)

Magnesia is the dopant of this type, usually using (8–10 mol%) of MgO to provide a biphasic composition of precipitates of tetragonal intragranular zirconia within the cubic matrix of stabilized zirconia through the introduction of magnesium cations within the tetragonal or cubic lattice of zirconia [9, 21]. Although (Mg-PSZ) exhibits transformation toughening with good chemical and thermal resistance properties, this material had precluded from several biomedical applications, it is considered an unstable material; due to its inferior properties resulting from the large grain size, which in turn increases the residual porosity and makes it susceptible to wear. Thus, it requires high sintering temperatures using special equipment for heating. Moreover, (Mg-PSZ) is almost impossible to be obtained purely free from alumina and silica, which diminishes magnesium content and allows ($t \rightarrow m$) transformation [22–24]. The incorporation of 8 mol% (Mg-PSZ) was confirmed to enhance the fracture toughness of (3Y-TZP) [25]. The mechanical properties of (Mg-PSZ) can be controlled by treatment with isothermal heat, controlled cooling followed by temperature-controlled sub-eutectoid aging [26]. Spark plasma sintering before sub-eutectoid aging was found to contribute to finetuning (Mg-PSZ) properties [27]. Several methods were used for (Mg-PSZ) preparation, such as solid-state reaction, sol-gel, electrospinning, precipitation, microwave, and sugar techniques [28]. *In vitro* studies demonstrated the enhancement of the biocompatibility of (Mg-PSZ) when coated with functionally graded bioactive glass, such results promote the fixation of this type of stabilized zirconia-based dental parts [29].

3. Investigation of zirconia in dental restorations

3.1 Zirconia dental posts

If the root of the dentine is not healthy and cannot serve teeth stability, a dental post is an option for treatment of the readily existing tooth as an alternative to the root to support and strengthen the teeth *via* retention of the restorations. Stabilized zirconia ceramics were introduced in dental posts for the first time in 1995 with comparable mechanical properties to other metal-based posts [30]. Many limitations in metal-based and all ceramic restorations encourage the employment of zirconia in dental posts. Corrosion activity of some metal-based posts leads to sensitization, unfavorable metallic taste, oral burnings, and pains. Furthermore, the esthetic concern has been a critical factor in recent years since metal posts usually do not fulfill this concern. This is clear when employed in anterior restorations, where unfavorable metallic grayish-blue discoloration due to the complete opacity of metals could affect the root and gingiva.

Such health and esthetic concerns dictate the development of translucent white dental posts with high chemical stability and biocompatibility with very low toxicity besides their mechanical properties. Zirconia is the favored efficient choice and confirmed as a material that attained high clinical success rates with strength comparable to all ceramic posts. **Figure 3** represents the root translucency difference between

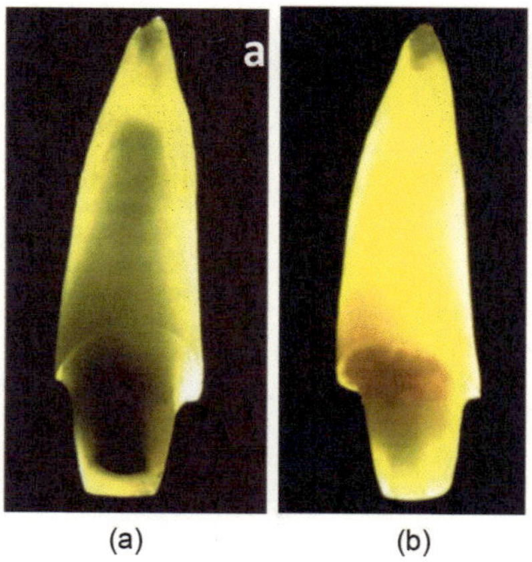

Figure 3.
Translucency of titanium (a) vs. TZP based posts (b) with composite build-up [30].

titanium-based versus TZP-based posts with composite build-up. To prevent the most severe problem, restorations fractures, the root should uniformly distribute the stress caused by the occlusal load; this could be achieved through the engagement of materials of lowered Young's modulus analogous to that of dentine.

Upon studying the features and disadvantages of zirconia-based dental posts, some drawbacks limit their application. Retreatment of the zirconia posts in dentine is complex since it is nearly impossible to remove or grind away the posts from the root canal; likewise, the root temperature will increase if ultrasonic vibration is used to remove fractured posts. Moreover, some studies proved the poor resin bonding of zirconia posts after dynamic loading. Dental resin is a bonding material for integration between the restoration parts [31]. Also, the relatively high Young's modulus and stiffness of zirconia promote zirconia posts to transfer the applied stress to the surrounding less rigid tooth, resulting in root fractures [32, 33].

3.2 Zirconia dental implants

The dental implants are similar to posts in their function as an alternative tooth root, whilst the implant is a foundation for a missing tooth and acts as a base for other prosthetic components for a totally-artificial tooth in the mandible or maxilla, as shown in **Figure 4(a)**. This foundation should conduct a robust functional and pleasant esthetic role. In the current century, zirconia-based implants proved to be an efficient alternative to other prominent metal-based implants used widely in the past decades, including alumina and titanium alloys.

Despite their high success rates, titanium alloys in dental implants had some unfavorable aspects that limited their applications, including the esthetic requirements and high wear properties, resulting in metallic particles causing allergic and toxic consequences. Alumina-based implants were limited with their high fracture susceptibility and inferior osseointegration. Stabilized zirconia, and (Y-TZP) implants in particular, are favored for their high flexural strength, tooth-like color, low-temperature conductivity, and masticatory forces bearing.

Several studies proved the biocompatibility of zirconia implants which represent a critical property as a component placed directly in contact with alveolar bone and connective tissues. The low toxicity of such implants was also confirmed

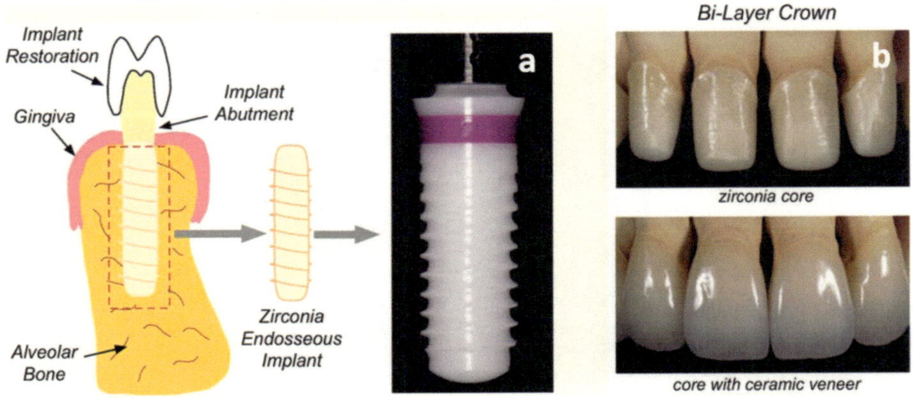

Figure 4.
Elucidative image of endosseous implant, abutment, and restoration (a). Bi-layer crowns of zirconia cores and translucent veneers (b) [8].

when tested using the immune system cells, this is supported by zirconia's low plaque affinity, avoiding surrounding tissues' inflammatory risks, and lower bacterial colonization compared to titanium implants. Zirconia osseointegration was directly influenced by surface roughness, which can be controlled *via* manufacturing processes, other than the machining process such as coating, sandblasting, and acid etching [8, 34].

3.3 Zirconia dental abutments

Zirconia abutment, the connection between the implant and the prosthesis as illustrated in **Figure 4(a)**, was introduced in 1996 [35]. It was developed to overcome titanium abutment esthetic limitations, with the maintenance of its mechanical strength. Zirconia could be sintered in titanium abutments to form titanium-reinforced zirconia with improved properties.

Previously, one-piece implants were designed without the introduction of the abutment as a connection, for concerns related to wear, fitting precision, and the colonization potential of bacteria and microbes in the micro-gap between the abutment and the implant from one side and the prosthetic (crown or bridge) from the other side. Such a one-piece implant system limited the versatility of the prosthesis compatibility with the implants' position. Furthermore, preparation of this type and grinding, in particular, may trigger ($t \rightarrow m$) transformation. Thus, a two-piece system was developed, including an implant and a separated abutment. This combination proved to achieve better osseointegration when the implant is not protruding, furthermore, better fracture resistance would result if the stress distributes over the two separate components.

An important point to be taken into account is that when a zirconia implant is fitted directly to the zirconia abutment, restoration fracture is already feasible; due to its inherent strength. Durable, nontoxic, and good adherence bonding cement is mandatory. Many types of cement were introduced, such as zinc phosphate, zinc polycarboxylate, glass, and resin ionomer. The last type is well-known for its ease of use and high retention [31]. Zirconia abutments are favored over the brittle alumina ones, on the other hand, zirconia abutments are more susceptible to fractures when compared to titanium-reinforced zirconia [12, 33, 36–38].

3.4 Zirconia in dental prosthesis

Dental prosthetics, including single prosthetic crowns (fixed dental prosthesis), bridges (multiple joined prosthesis), and dentures (Full jaw prosthesis), are used as an alternative for missing dentine. It is a major part of the restoration in which functional and esthetic considerations must be satisfied; due to its exposure to harsh conditions and bearing direct pressure and masticatory loads. Moreover, as an external part of the restoration, the esthetic appearance is a priority, especially for the anterior dentine [12]. Among the versatile types of ceramic restorations; zirconia and lithium disilicate-based prostheses showed intrinsic promising functional capability as in **Figure 5**. Zirconia prosthetics are preferred for their superior mechanical properties and have become a popular trend among dentists as a material that fulfills the prementioned requirements. CAD/CAM technologies' advancement in zirconia production also facilitates its selection as the approved type.

Prosthesis preparation is a critical factor for its fitting with both the foundation and the veneering components [9]. Transformation toughening is crucial in making its mechanical properties exceed other ceramic prosthetics. (3Y-TZP) is widely preferred for its high flexural strength and fracture toughness. Versatile types could be obtained with different properties *via* variations of stabilized zirconia concentration, grain size, sintering conditions, coloration technique, and surface processing and treatment [39].

Despite the favored clinical performance of zirconia crowns and other prosthetics, it has some drawbacks in bi-layer prosthesis, as shown in **Figure 4**(b), which consists of veneering porcelain that works as a translucent mask over the opaque white zirconia, mainly for esthetic concerns, and particularly in anterior regions. The veneering porcelain in bi-layer prosthetics is susceptible to chipping and aging problems in the presence of water. Chipping is the cohesive frailer between the zirconia substructure and the veneer.

The veneering problem is attributed to many reasons, including a mismatch of thermal expansion coefficient between the two layers, vacant sites due to porosity,

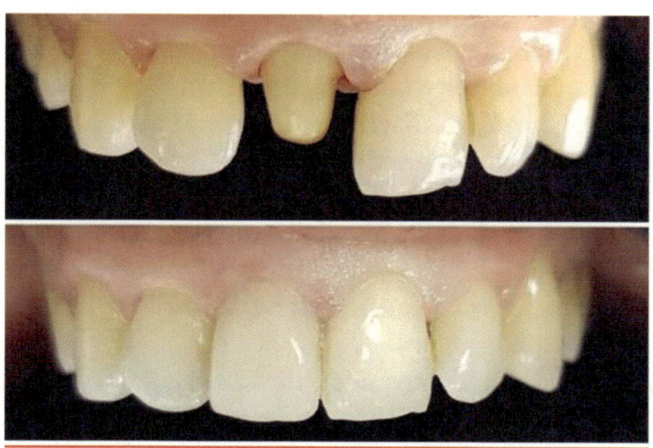

Figure 5.
Zirconia and lithium disilicate-based prostheses.

lower fracture toughness of the veneer structure, overloading, and inappropriate design of zirconia frameworks with insufficient support of porcelain over the framework. Chipping could be solved in some cases simply by polishing the rough sites, using composite resin for fracture treatment, or total replacement in some complicated cases [40]. Monolithic zirconia recently is preferred, in which a one-piece component is introduced instead of two separated components core and veneer.

3.5 Zirconia in esthetic brackets

Several types of brackets have been used for orthodontic treatment, such as stainless steel, titanium composites, and alumina. Zirconia brackets were introduced as the type of greatest toughness among other ceramic-based types. Its frictional coefficient is lower than that of alumina, which is a feature. Nevertheless, zirconia's high opacity inhibits the esthetic appearance, especially with the growing trends for translucent brackets [41]. Recent advances work on developing the zirconia bracket's structure to enhance its properties and shorten the treatment period. A treatment using zirconia brackets fabricated in a specific process with three slots within its structure, , connected with nickel-titanium arch-wires was successful, in which the tooth movement occurs at a higher rate. A shorter treatment period was observed when compared to edge-wise appliances [42].

3.6 Zirconia properties

Understanding zirconia properties helps in a better conception of its performance and applications. One of the important phenomena in zirconia is low-temperature degradation (LTD) or aging described previously in the second section. Other properties will be described in this section:

4. Biocompatibility

Biocompatibility could be described as the ability of a biomaterial or medical component to carry out a medical therapeutic function usually for long-term contact with human body tissues, with relatively no pathogenic harmful side effects, such as inflammation, cytotoxicity, allergy, or carcinogenic effects [43]. Many researchers tested zirconia biocompatibility in dental restoration. It is proven that partially stabilized zirconia performed favorable initial fibroblast adhesion on its surface, which enhances the growth of the connective tissues when compared to titanium, polystyrene, and fully stabilized zirconia. Increasing the concentration of yttria in partially stabilized zirconia produces fully stabilized zirconia with a rougher surface which affects the adhesion behavior [44]. *In vivo* and *in vitro* studies proved that zirconia is an osseoconductive biomaterial in addition to being safe and stable without reported carcinogenic or mutagenic impacts [45]. Some studies worked on modifying zirconia surfaces, such as incorporating calcium ions, which enhanced the biocompatibility of such restorations without affecting their mechanical properties [46]. In general, no cytotoxic or damaging side reactions manifest from the introduction of zirconia in dental applications, in addition to the good bone response with relatively no inflammation or bacterial growth within acceptable levels [47].

4.1 Optical properties and translucency

Translucency is an intermediate property between transparency and total opaque. Translucent material allows the transmission of light with dispersion which obstacles a clear seen of objects through it [48]. It is a requirement for esthetic aspects in dental restorations. Zirconia opacity is a drawback despite its naturally white color, but zirconia's translucency is required since color compatibility is essential. This material also showed relatively high X-Ray opacity obstructing diagnosis [47]. Therefore, efforts worked on the enhancement of zirconia translucency in several ways and attempted to control its optical properties and refractive index, to tong the top choice for restorative components. Many factors could affect zirconia's translucency, such as impurities, porosity, grain size, restoration thickness [49], and processing conditions. Controlling these factors could enhance translucency [50].

4.2 Radioactivity

Zirconia was found to contain small portions of radionuclides from uranium, radium, and thorium series type. It is possible to obtain zirconia of radioactivity within acceptable limits *via* purification processes [9]. To understand the effect of composition on zirconia radioactivity, an *in vitro* study represents the radioactivity of three common types of dental zirconia composites using gamma spectrometry, Vita In-Ceram YZ, Zirkonzahn, and Zirkonzahn Prettau (**Table 3**). Zirkonzahn Prettaud had the highest radioactivity as **Figure 6** represents. Even though, all results were within acceptable limits of 1000 Bq/kg according to the International Atomic Energy Agency (IAEA) [51].

4.3 Wear behavior

Surface roughness influences the abrasion of material and the wear with the adjacent teeth. Enamel wear is affected directly by the surface roughness and surface microstructures, in addition to the surrounding environmental conditions. The surface roughness could be raised by grinding and decreased by polishing [52].

Vita In-Ceram YZ		Zirkonzahn		Zirkonzahn Prettau	
Components	**Weight (%)**	**Components**	**Weight (%)**	**Components**	**Weight (%)**
ZrO_2	90.9–94.5	ZrO_2	Main	ZrO_2	Main
Y_2O_3	4–6	Y_2O_3	4–6	Y_2O_3	4–6
HfO_2	1.5–2.5	Al_2O_3	<1	Al_2O_3	<1
Al_2O_3	0–0.3	Na_2O	max. 0.04	Na_2O	max. 0.04
Fe_2O_3	0–0.3	SiO_2	max. 0.02	SiO_2	max. 0.02
—	—	Fe_2O_3	max. 0.01	Fe_2O_3	max. 0.01

Table 3.
Chemical composition of three common zirconia types [51].

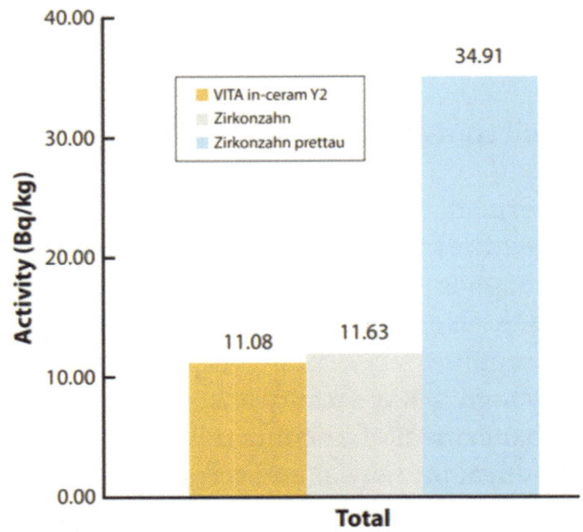

Figure 6.
Total radioactivity for each type [51].

5. Manufacturing of dental zirconia

Zirconia-based dental restorations nowadays are designed and manufactured using CAD/CAM technology. Two techniques are mainly used for the manufacturing process; soft and hard machining. Soft machining is based on the milling process of pre-sintered blanks produced by cold isostatic pressing of compact zirconia powder in the presence of a binder. Later on, those blanks will be fully sintered. High temperature is utilized for the sintering step. This technique is usually used for (3Y-TZP) manufacturing.

On the other hand, the hard machining technique uses fully sintered stabilized zirconia blocks. The pre-sintered zirconia is obtained at a temperature of less than 1500°C to maintain the required density. Then, hot isostatic pressing is obtained at temperatures between 1400 and 1500°C under inert conditions and high pressure to maintain high density, hardness, and homogeneity of the blocks. A special milling system then performs machining to obtain the required dimensions. The milling system has to be strong and robust because the fully sintered zirconia is very hard with low machinability [9, 15].

6. Surface modifications for zirconia dental implants

Many factors could affect the nature and properties of the fabricated zirconia and its quality and biocompatibility, such as its chemical composition, morphology, and surface roughness [53]. In this section, the main surface treatment methods of zirconia-based dental restorations are discussed, which are found to have a direct influence on its osseointegration as well as the mechanical properties. Various classifications for the treatment methods were used; herein the classification of these methods is presented within three categories; physical treatment, chemical treatment, and coating processes.

6.1 Physical treatment

6.1.1 Surface sandblasting

It is a surface abrasion process in which the pressure of the compressed air ejects particles strongly to acquire a surface with micro-roughness. In this process, the abrasion is performed as a homogeneous and anisotropic abrasion on the hard material [54]. Many studies confirm the advantageous role of increasing zirconia-based implant roughness with sandblasting on the amount of integration and contact with the bone tissues. The roughness is detected by a parameter known as bone-implant contact (BIC), in addition to improving the strength of the bond, detected by the removal torque parameter (RTQ) [55].

Sandblasting of zirconia with alumina was confirmed to increase its surface roughness. Thus, highly efficient initial adhesion of human osteoblasts cells was achieved [56]. However, this process has some drawbacks; sandblasting with alumina will affect the elemental chemical composition of the treated surface, in which alumina act as a contaminant. The problem could be solved by utilizing the acid etching treatment [54]. Moreover, LTD is susceptible to occurring with increased sandblasting pressure due to the impact of the mechanical forces on the surface [57].

6.1.2 Laser

Laser is a promising treatment for zirconia surface to improve its osseointegration. Unlike sandblasting, zirconia phase transformation in laser treatment is uncommon and surface contamination is avoided since there is no direct contact between the laser and the surface [54]. This type of treatment is fast, easy to operate, clean, and highly accurate; it promotes micro-grooved implant surface and thus enhances the surface roughness. Laser treatment promotes surface properties changes, including roughness, topography, and wettability. This treatment type was confirmed to improve surface micro and nano-scaled roughness; this resulted in enhanced wettability, which directly influences biocompatibility, cell adhesion, and proliferation.

The properties of the treated surface depend on the irradiation conditions, which could be optimized by controlling the irradiation frequency, intensity, and time [58]. The femtosecond laser established a consistent roughness between the surface and the bonding resin, increasing the bond strength. A fiber laser could achieve wider adequate grooves; this advances the (BIC) and (RTQ) parameters [4, 55].

6.1.3 Ultraviolet (UV) light

As mentioned before, wettability is correlated with biocompatibility and integration with bone tissues. Many studies proved that surface treatment with UV light increases wettability by lowering the surface contact angle below 20° to obtain a hydrophilic and even super-hydrophilic surface. The hydrophilicity surface character is mainly obtained due to decreasing the superficial hydrocarbons *via* UV light [59], and increasing the oxygen character. When the treated hydrophilic oxide binds to water in the wet environment near the tissues, the surface will be in the form of hydroxylated oxide, which induces surface reactivity toward the surrounding amino acids and proteins [54]. Thus, zirconia treated with UV radiation exhibits very good osteoblast response and proliferation. However, the influence of UV light on zirconia aging still needs further effort and study. Many studies indicated that UV treatment reduces the ability of zirconia to age, while some controversial studies found that UV light triggers crystalline transformation [55].

6.2 Chemical treatment

6.2.1 Acid etching

Acids, such as HF, HNO_3, or H_2SO_4 usually used for zirconia surface treatment in the acid etching process to increase its roughness homogeneously, even in the case of irregular surfaces, without destroying its morphology. Acid etching is effective in overcoming the prementioned sandblasting contamination problem by removing excessive residues. Thus, acid etching is usually employed in conjunction with a previous surface sandblasting. Sandblasting of zirconia implants using large grits followed by surface etching with a strong acid provides SLA implants, an abbreviation for sandblasted, large grit, acid-etched implant surface, with an increased surface area, promoting surface bio-adhesion [60, 61]. A comparative animal study tested three different types of sandblasted zirconia implants, the three types are acid-etched (SLA), alkaline-etched, and sandblasted zirconia without etching. The results indicated the highest BIC values were attributed to the acid-etched SLA type, while the alkaline-etched implants achieved the lowest BIC values [62].

6.2.2 Electrochemical treatment

Despite the nonconductive character of zirconia, many electrochemical treatment methods were proven to enhance the properties of zirconia dental restorations. For example, electrochemical deoxidation of zirconia (ECD) improved its biocompatibility by developing a micro-porous surface and thus lowering the contact angle. This method is known to enhance surface wettability through oxygen removal from the surface of the solid zirconia using molten salt electrolysis [63].

Recent electrochemical techniques focused on producing nanostructures on the surface of zirconia such as nanotubes [64]. The electrochemical anodization technique (EA) is extensively used for the fabrication of zirconia nanopores or nanotubes since it is a cost-effective technique and can control the physical and chemical properties of the prepared nanostructures. Zirconia nanotubes were found to promote the stability of zirconia implants with better initial cell adhesion. The fabricated nanotubes could be modified *via* annealing to improve their corrosion resistance [65].

6.3 Coating

Several types of bioactive coatings are employed to augment the function of the osteoblasts. Some of these coating materials, such as calcium phosphate, polydopamine, bioactive glass, and biomolecular coatings are discussed in this section.

6.3.1 Calcium phosphate

Calcium phosphate is a mineral component in bones; this critical point accelerates and supports the osseointegration when zirconia implants are coated with these compounds. Furthermore, calcium deposition on such coated zirconia implants and the adhesion of proteins will be enhanced [61]. This type of coating is affected by the properties of the utilized compound since several compounds are implicated as members of the calcium phosphate family, such as β-tricalcium phosphate (β-TCP) and the most stable hydroxyapatite (HA) with the chemical formula $Ca_{10}(PO_4)_6(OH)_2$.

Several methods are employed for coatings. The plasma spraying method is favored over sol-gel, wet powder spraying, and aerosol deposition techniques since it is inexpensive and its deposition rate is high, but not suitable for complex morphologies [55, 66]. However, calcium phosphate bonding on the zirconia surface is relatively weak; several studies attempted to reinforce bonding in different ways, such as laser treatment before coating, or the application of coating along with (HA) and hydrothermal sintering after the coating process [65].

6.3.2 Polydopamine (PDA)

The good adhesion properties of marine mussels' proteins attracted the interest to take advantage of the components in these proteins. Upon studying this property, the unusual amino acid 3,4-dihydroxy-L-phenylalanine referred to as (L-DOPA) is confirmed as the liable component. Dopamine is considered an (L-DOPA) precursor [67]. The introduction of dopamine and polydopamine was approved to augment the bioactivity of the zirconia surface and enhance cell adhesion over the surface. The enhancement is attributed to the easy adsorption processes taking place on the surfaces due to the strong anchoring of the catechol functionality [68]. This type of coating also reduces bacterial adhesion on the coated surface and improves antimicrobial activity, which aids in the rapid healing and regeneration of the soft tissues surrounding the dental implants. Moreover, PDA coating is simple and nontoxic [67].

6.3.3 Coating with biomacromolecules

Arginine-glycine-aspartate, referred to as RGD is a tripeptide biomolecule known for its proteins' adhesion properties. It is found that RGD plays a key role in the adhesion of osteogenic cells and thus holds significant promise to achieve efficient bioactivity if incorporated with dental implants [69]. Zirconia surface coating with RGD was accomplished through the immobilization of the coating material on the zirconia implant surface by a process called surface biofunctionalization or biomimetic modification. This process aid in the enhancement of its biological character. RGD coated (Y-TZP). The ability of successful chemical bonding between RGD and (Y-TZP) was proven, and this combination performed better biocompatibility [54, 70].

6.3.4 Bioactive glass

Bio-glass, a composition of sodium, calcium, silicon, and phosphorous oxides, is considered a bioactive material, which promotes its introduction in zirconia surface coating. It forms hydroxyapatite between the implant's surface and biological tissues when present within a biological environment [55]. This type of coating is proposed to enhance the biocompatibility of the implants and reduce the healing period. However, the success of these attempts was limited because of several drawbacks in bio-glass. Its mechanical properties are insufficient, making it a fragile component. Furthermore, its thermal expansion coefficient is relatively high, so it is unsuitable for zirconia thermal coating, and cracking of the coating is susceptible [61].

7. Challenges and future developments of zirconia application

To overcome the dental veneer drawbacks, chipping, and delamination, in addition to thickness considerations, attention has focused on the development and further modifications of the monolithic zirconia. To enhance monolithic zirconia's properties, chemical dopants are utilized. Increasing the yttria content with a lowering in alumina content and the incorporation of (0.2 mol%) La_2O as a dopant resulted in improved (3Y-TZP) translucency and aging resistance enhanced. Increased yttria concentration to (5Y-TZP) resulted in superior translucency and aging resistance properties, but the toughness was scarified. Zirconia opacity is still a drawback that prevents its participation in the anterior sites. Dopants incorporation is considered a promising technique. The opacity of stabilized zirconia could be attributed to the light scattering performed by its grain boundaries and microstructural defects. Alumina was found to increase light scattering, thus its content in zirconia should decrease, but not eliminate, to a level below (0.25 wt%). The transmittance was enhanced when the grain size of zirconia was reduced to a diameter of less than 100 nm [71].

As known, nanotechnology is spreading every day through most of the manufacturing processes and applications; due to its distinct properties and the superior development of the material's character. For example, nano-powders with a regulated composition of stabilizing material or additives could be used for the fabrication of zirconia and has found to improve grains development during sintering and decreases porosity. But for this technique, new processing methods should be developed; since the normal fabrication method used for processing zirconia with nano-powders was very hard [52]. Some studies confirmed the promising properties and positive role of alumina-zirconia nanocomposite with better toughness and high capability for aging resistance in addition to crack propagation inhibition, which increases its reliability in medical applications [33].

The future of stabilized zirconia in the material science field is promising and requires intensive efforts and searching for new strategies to withstand the challenge of enhancing the esthetic properties while maintaining the mechanical properties of zirconia at the same time. This is a game of advancements and compromises.

8. Conclusion

Zirconia has attracted significant attention recently, especially in dental applications for mechanical and esthetic considerations. Herein, we demonstrate a general view of zirconium, zircon, and zirconia. This chapter discussed the main types of stabilized zirconia incorporated in dental restorations: Y-TZP, ZTA, and Mg-PSZ. Zirconia-based dental restorations, including posts, implants, abutments, fixed denture prostheses, and orthodontic brackets, were explained. Then, an illustration of some of the substantial zirconia properties that directly affect its mechanical and esthetic properties, such as (LTD) or aging, zirconia biocompatibility, optical properties, translucency, and radioactivity. Surface modifications of dental zirconia are also presented. The physical treatment techniques: sandblasting, laser, and UV light, in addition to the chemical treatment, including acid etching and electrochemical treatment, were discussed. Different coatings utilizing calcium phosphate, polydopamine, bio-macromolecules, and bioactive glass were introduced. Finally, we demonstrated recent developments, challenges, and directions for future research to enhance the survival rates of different zirconia-based dental restorations.

Conflict of interest

The authors declare no conflict of interest.

Author details

Ghassan Albarghouti* and Haneen Sadi
Birzeit University, Ramallah, Palestine

*Address all correspondence to: gdaghra@birzeit.edu

References

[1] Lide DR. CRC Handbook of Chemistry and Physics [Internet]. Boca Taton: Taylor and Francis;FL: CRC Press; 2005

[2] Smirnova DE, Starikov SV, Gordeev IS. Evaluation of the structure and properties for the high-temperature phase of zirconium from the atomistic simulations. Computational Materials Science. 2018;**152**:51-59

[3] Rodaev VV, Zhigachev AO, Tyurin AI, Razlivalova SS, Korenkov VV, Golovin YI. An engineering zirconia ceramic made of baddeleyite. Materials. 2021;**14**(16):4676

[4] Nielsen RH, Schlewitz JH, Nielsen H. Updated by StaffZirconium and zirconium compounds. In: Kirk-Othmer Encyclopedia of Chemical Technology. 2013

[5] Albarghouti G, Rayyan S. General method for the synthesis of substituted Cyclopentenones via α-Borylzirconacyclopentene intermediates. Organic Preparations and Procedures International. 2020;**52**(1):1-8

[6] Vagkopoulou T, Koidis P, Strub JR, Dent M. Zirconia in dentistry: Part 1. Discovering the nature of an upcoming bioceramic. European Journal of Esthetic Dentistry. 2009;**4**:130-151

[7] Maziero Volpato CA, Altoe Garbelotto LGD, Celso M, Bondioli F. Application of Zirconia in Dentistry: Biological, Mechanical and Optical Considerations [Internet]. Advances in Ceramics - Electric and Magnetic Ceramics, Bioceramics, Ceramics and Environment. InTech; 2011. DOI: 10.5772/21630

[8] Chen Y-W, Moussi J, Drury JL, Wataha JC. Zirconia in biomedical applications. Expert Review of Medical Devices. 16 Sep 2016;**13**(10):945-963. DOI: 10.1080/17434440.2016.1230017

[9] Asharaf S, Suma A, Deivanai M, Mani R. Zirconia: Properties and application — A review. Pakistan Oral & Dental Journal. 2014;**34**(1):178-183

[10] Tanaka H, Maeda T, Narikiyo H, Morimoto T. Mechanical properties of partially stabilized zirconia for dental applications. Journal of Asian Ceramic Societies. 2019;**7**(4):460-468

[11] Kelly JR, Denry I. Stabilized zirconia as a structural ceramic: An overview. Dental Materials. 2008;**24**:289-298

[12] Grech J, Antunes E. Zirconia in dental prosthetics: A literature review. Journal of Materials Research and Technology. 2019;**8**:4956-4964

[13] Khaskhoussi A, Calabrese L, Currò M, Ientile R, Bouaziz J, Proverbio E. Effect of the compositions on the biocompatibility of new alumina-zirconia-titania dental ceramic composites. Materials. 2020;**13**(6):1374

[14] Nakai H, Inokoshi M, Nozaki K, Komatsu K, Kamijo S, Liu H, et al. Additively manufactured zirconia for dental applications. Materials. 2021;**14**(13):3694

[15] Zarone F, Russo S, Sorrentino R. From porcelain-fused-to-metal to zirconia: Clinical and experimental considerations. Dental Materials. 2011;**27**(1):83-96

[16] Sarker S, Mumu HT, Al-Amin M, Zahangir Alam M, Gafur MA. Impacts of inclusion of additives on physical, microstructural, and mechanical

properties of alumina and zirconia toughened alumina (ZTA) ceramic composite: A review. Materials Today: Proceedings. 2022;**62**:2892-2918

[17] De Aza AH, Chevalier J, Fantozzi G, Schehl M, Torrecillas R. Crack growth resistance of alumina, zirconia and zirconia toughened alumina ceramics for joint prostheses. Biomaterials. 2002;**23**:937-945

[18] Nevarez-Rascon A, Aguilar-Elguezabal A, Orrantia E, Bocanegra-Bernal MH. On the wide range of mechanical properties of ZTA and ATZ based dental ceramic composites by varying the Al2O3 and ZrO2 content. International Journal of Refractory Metals and Hard Materials. 2009;**27**(6):962-970

[19] Piconi C, Sprio S. Oxide bioceramic composites in orthopedics and dentistry. Journal of Composites Science. 2021;**5**:206

[20] Mussano F, Genova T, Munaron L, Faga MG, Carossa S. Ceramic biomaterials for dental implants: Current use and future perspectives. In: Dental Implantology and Biomaterial. London, UK: InTech; 2016

[21] Ispas A, Iosif L, Murariu-Măgureanu C, Craciun A, Constantiniuc M. Zirconia in dental medicine: a brief overview of its properties and processing techniques [Internet]. 2021. Available from: http://www.hvm.bioflux.com.ro/

[22] Denry I, Kelly JR. State of the art of zirconia for dental applications. Dental Materials. 2008;**24**(3):299-307

[23] Hisbergues M, Vendeville S, Vendeville P. Review zirconia: Established facts and perspectives for a biomaterial in dental implantology. Journal of Biomedical Materials Research - Part B Applied Biomaterials. 2009;**88**:519-529

[24] Alhusainy A. Zirconia As Dental Restoration Material. The Journal of Pharmaceutical Negative Results. 2023;**13**:7714-7723

[25] Soylemez B, Sener E, Yurdakul A, Yurdakul H. Fracture toughness enhancement of yttria-stabilized tetragonal zirconia polycrystalline ceramics through magnesia-partially stabilized zirconia addition. Journal of Science: Advanced Materials and Devices. 2020;**5**(4):527-534

[26] Sengupta P, Bhattacharjee A, Maiti HS. Zirconia: A Unique Multifunctional Ceramic Material. Transactions of the Indian Institute of Metals. 2019;**72**(8):1981-1998

[27] Cho J, Yang B, Shen C, Wang H, Zhang X. Micromechanical properties and microstructure evolution of magnesia partially stabilized zirconia prepared by spark plasma sintering. Journal of the European Ceramic Society. 2023;**43**(3):1098-1107

[28] Wahyudi K, Maryani E, Arifiadi F, Rostika A, Yusuf D, Manullang RJ, et al. Low-calcination temperatures of magnesia partially stabilized zirconia (Mg-PSZ) nanoparticles derived from local zirconium silicates. Materials Research Express. 2021;**8**(4):045022

[29] Rahaman MN, Li Y, Bal BS, Huang W. Functionally graded bioactive glass coating on magnesia partially stabilized zirconia (Mg-PSZ) for enhanced biocompatibility. Journal of Materials Science. Materials in Medicine. 2008;**19**(6):2325-2333

[30] Meyenberg KH, Lüthy H, Schärer P. Zirconia posts: A

new all-ceramic concept for nonvital abutment teeth. Journal of Esthetic and Restorative Dentistry. 1995;**7**(2):73-80

[31] Della BA, Pecho OE, Alessandretti R. Zirconia as a dental biomaterial. Materials. 2015;**8**:4978-4991

[32] Özkurt Z, Işeri U, Kazazoğlu E. Zirconia ceramic post systems: A literature review and a case report. Dental Materials Journal. 2010;**29**:233-245

[33] Madfa AA, Al-Sanabani FA, Al-Qudami NH, Al-Sanabani JS, Amran AG. Use of zirconia in dentistry: An overview. The Open Biomaterials Journal. 2014;**5**:1-9

[34] Wennerberg A, Jimbo R, Albrektsson T. Implant Surfaces and their Biological and Clinical Impact. Berlin/ Heidelberg, Germany: Springer; 2015

[35] Silva NRFA, Sailer I, Zhang Y, Coelho PG, Guess PC, Zembic A, et al. Performance of zirconia for dental healthcare. Materials. 2010;**3**:863-896

[36] Butz F, Heydecke G, Okutan M, Strub JR. Survival rate, fracture strength and failure mode of ceramic implant abutments after chewing simulation. Journal of Oral Rehabilitation. 2005;**32**(11):838-843

[37] Aras MA, Modgi CM. Zirconia abutments in implant dentistry. International Journal of Oral Implantology & Clinical Research. 2012;**3**(1):39-42

[38] Gomes AL, Montero J. Zirconia implant abutments: A review. Medicina Oral, Patologia Oral, Cirugia Bucal. 2011;**16**(1):e50-e55

[39] Guazzato M, Quach L, Albakry M, Swain M, v. Influence of surface and heat treatments on the flexural strength of Y-TZP dental ceramic. Journal of Dentistry. 2005;**33**(1):9-18

[40] Al-Amleh B, Lyons K, Swain M. Clinical trials in zirconia: A systematic review. Journal of Oral Rehabilitation. 2010;**37**:641-652

[41] Keith O, Kusy RP, Whitley JQ, Hill C. Zirconia brackets: An evaluation of morphology and coefficients of friction. American Journal of Orthodontics and Dentofacial Orthopedics. 1994;**106**(6):605-614

[42] Maki K, Futaki K, Tanabe S, Takahashi M, Ichikawa Y, Yamaguchi T. Applicative characteristics of a new zirconia bracket with multiple slots. International Journal of Dentistry. 2016;**2016**:4348325

[43] Williams DF. On the mechanisms of biocompatibility. Biomaterials. 2008;**29**(20):2941-2953

[44] Takamori ER, Cruz R, Gonçalvez F, Zanetti RV, Zanetti A, Granjeiro JM. Effect of roughness of zirconia and titanium on fibroblast adhesion. Artificial Organs. 2008;**32**(4):305-309

[45] Harianawala H, Kheur M, Bal A. Biocompatibility of zirconia. Journal of Advanced Medical and Dental Sciences Research. 2016;**4**(3):1-5

[46] Sasaki K, Hayashi T, Asakura M, Ando M, Kawai T, Ban S. Improving biocompatibility of zirconia surface by incorporating ca ions. Dental Materials Journal. 2015;**34**(3):336-344

[47] Nistor L, Grădinaru M, Rîcă R, Mărășescu P, Stan M, Manolea H, et al. Zirconia use in dentistry - manufacturing and properties. Current Health Sciences Journal. 2019;**45**(1):28-35

[48] Turgut S, Bagis B, Turkaslan SS, Bagis YH. Effect of ultraviolet aging on

translucency of resin-cemented ceramic veneers: An In vitro study. Journal of Prosthodontics. 2014;**23**(1):39-44

[49] Abdelbary O, Wahsh M, Sherif A, Salah T. Effect of accelerated aging on translucency of monolithic zirconia. Future Dental Journal. 2016;**2**(2):65-69

[50] Manziuc MM, Gasparik C, Negucioiu M, Constantiniuc M, Burde A, Vlas I, et al. Optical properties of translucent zirconia: A review of the literature. EuroBiotech Journal. 2019;**3**(1):45-51

[51] Brizuela-Velasco A, Chento-Valiente Y, Chávarri-Prado D, Pérez-Pevida E, Diéguez-Pereira M. Zirconia and radioactivity: An in vitro study to establish the presence of radionuclides in dental zirconia. Journal of Prosthetic Dentistry. 2021;**126**(1):115-118

[52] Alqutaibi AY, Ghulam O, Krsoum M, Binmahmoud S, Taher H, Elmalky W, et al. Revolution of current dental zirconia: A comprehensive review. Molecules. 2022;**27**:1699

[53] Huang YS, Huang HH. Effects of clinical dental implant abutment materials and their surface characteristics on initial bacterial adhesion. Rare Metals. 2019;**38**:512-519

[54] Karthigeyan S, Ravindran A, Bhat R, Nageshwarao M, Murugesan S, Angamuthu V. Surface Modification Techniques for Zirconia-Based Bioceramics: A Review. Journal of Pharmacy and Bioallied Sciences. 2019;**11**:S131-S134

[55] Sun L, Hong G. Surface modifications for zirconia dental implants: A review. Frontiers in Dental Medicine. 2021;**2**:733242

[56] WMS AQ, Schille C, Spintzyk S, MSA AQ, Engel E, Geis-Gerstorfer J, et al. Effect of surface modification of zirconia on cell adhesion, metabolic activity and proliferation of human osteoblasts. Biomedizinische Technik. 2017;**62**(1):75-87

[57] Su N, Yue L, Liao Y, Liu W, Zhang H, Li X, et al. The effect of various sandblasting conditions on surface changes of dental zirconia and shear bond strength between zirconia core and indirect composite resin. Journal of Advanced Prosthodontics. 2015;**7**(3):214-223

[58] Cunha W, Carvalho O, Henriques B, Silva FS, Özcan M, Souza JCM. Surface modification of zirconia dental implants by laser texturing. Lasers in Medical Science. 2022;**37**:77-93

[59] Brezavšček M, Fawzy A, Bächle M, Tuna T, Fischer J, Att W. The effect of UV treatment on the osteoconductive capacity of zirconia-based materials. Materials. 2016;**9**(12):958

[60] Kligman S, Ren Z, Chung CH, Perillo MA, Chang YC, Koo H, et al. The impact of dental implant surface modifications on osseointegration and biofilm formation. Journal of Clinical Medicine. 2021;**10**:1641

[61] Schünemann FH, Galárraga-Vinueza ME, Magini R, Fredel M, Silva F, Souza JCM, et al. Zirconia surface modifications for implant dentistry. Materials Science and Engineering C. 2019;**98**:1294-1305

[62] Saulacic N, Erdösi R, Bosshardt DD, Gruber R, Buser D. Acid and alkaline etching of sandblasted zirconia implants: A Histomorphometric study in miniature pigs. Clinical Implant Dentistry and Related Research. 2014;**16**(3):313-322

[63] Liu J, Hong G, Wu YH, Endo K, Han JM, Kumamoto H, et al. A novel

method of surface modification by electrochemical deoxidation: Effect on surface characteristics and initial bioactivity of zirconia. Journal of Biomedical Materials Research. Part B, Applied Biomaterials. 2017;**105**(8):2641-2652

[64] Tsuchiya H, MacAk JM, Taveira L, Schmuki P. Fabrication and characterization of smooth high aspect ratio zirconia nanotubes. Chemical Physics Letters. 2005;**410**(4-6):188-191

[65] Chopra D, Jayasree A, Guo T, Gulati K, Ivanovski S. Advancing dental implants: Bioactive and therapeutic modifications of zirconia. Bioactive Materials. 2022;**13**:161-178

[66] Rao GN, Pampana S, Yarram A, MC SS, AV R, rao D. B malingeswara. Surface modifications of dental implants: An overview. International journal of. Dental Materials. 2019;**1**(1):17-24

[67] Liu M, Zhou J, Yang Y, Zheng M, Yang J, Tan J. Surface modification of zirconia with polydopamine to enhance fibroblast response and decrease bacterial activity in vitro: A potential technique for soft tissue engineering applications. Colloids and Surfaces. B, Biointerfaces. 2015;**136**:74-83

[68] Yu Y, Cui M, Lin T, Liu C, Jin X, Chen H, et al. Effect of zirconia surface modification using dopamine polymerisation on the shear bond strength of resin cement. European Journal of Oral Sciences. 2021;**129**(4):e12797

[69] Hsu SK, Hsu HC, Ho WF, Yao CH, Chang PL, Wu SC. Biomolecular modification of zirconia surfaces for enhanced biocompatibility. Thin Solid Films. 2014;**572**:91-98

[70] Soon G, Pingguan-Murphy B, Lai KW, Akbar SA. Review of zirconia-based bioceramic: Surface modification and cellular response. Ceramics International. 2016;**42**:12543-12555

[71] Zhang F, Inokoshi M, Batuk M, Hadermann J, Naert I, van Meerbeek B, et al. Strength, toughness and aging stability of highly-translucent Y-TZP ceramics for dental restorations. Dental Materials. 2016;**32**(12):e327-e337

Chapter 7

Perspective Chapter: The Application of Yttria-Stabilized Zirconia (YSZ)

Alexander Chee Hon Cheong and SivaKumar Sivanesan

Abstract

Pure zirconia will transform into different phases, which include monoclinic, tetragonal, and cubic, at different high temperature levels. Specific phases can be retained at room temperature by adding stabilizer and yttria is one of the most common stabilizers for zirconia, commonly formed yttria stabilizer zirconia (YSZ). To utilize YSZ in various industry applications, the amount of yttria and sintering temperature played a vital role. Thus far, YSZ has received a warm welcome in the industries of thermal barrier coating (TBC), solid oxide fuel cell (SOFC), and biomaterial. However, the limitations and challenges still occur, and this opens up the room and possibility of enhancing and improving the material properties of YSZ for a better performance in the mentioned area. This chapter explained the working principles of YSZ in the industries respectively and the research been conducted to improve the materials accordingly.

Keywords: zirconia, application, biomaterial, SOFC, TBC

1. Introduction

Zirconia (ZrO_2) possessed attractive material properties (mechanical, thermal, and electrical properties) when additional conditions are included. However, the material will encounter phase transformation at different levels of temperature [1]. As shown in **Figure 1**, the crystal structure of zirconia will transform from monoclinic to tetragonal and to cubic as the temperature is increasing, and the transformation is reversible.

By adding stabilizers to zirconia, it is one of the strategies to enhance the material properties of zirconia through retaining a specific phase, which the ceramics material did not come naturally at room temperature. Yttria (Y_2O_3) is one of the most common stabilizers for zirconia and formed yttria stabilizer zirconia (YSZ). The phase transition of YSZ is complicated but is well documented as the sintering temperature and the yttria concentration vary. The phase diagram, which is illustrated in **Figure 2**, clearly indicated the phase (or phases) of YSZ will be produced as the mole percentages (mol%) of yttria against the sintering temperature. For example, tetragonal (t) and cubic (c) will be produced with 5 mol% of yttria (5Y) and sintered at temperature 1000°C.

The Scanning Electron Microscope (SEM) samples images of different yttria amount is showed in **Figure 3**. Under the same level of close view observation, the images revealed the higher the amount of yttria, the higher the grain size of the

1170 °C 2370 °C 2715 °C

Monoclinic <----> Tetragonal <----> Cubic <----> Liquid

Figure 1.
The phase transformation of zirconia.

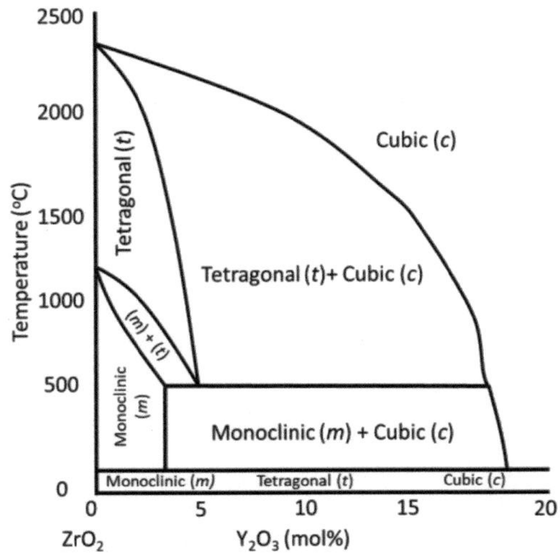

Figure 2.
Phase diagram of YSZ.

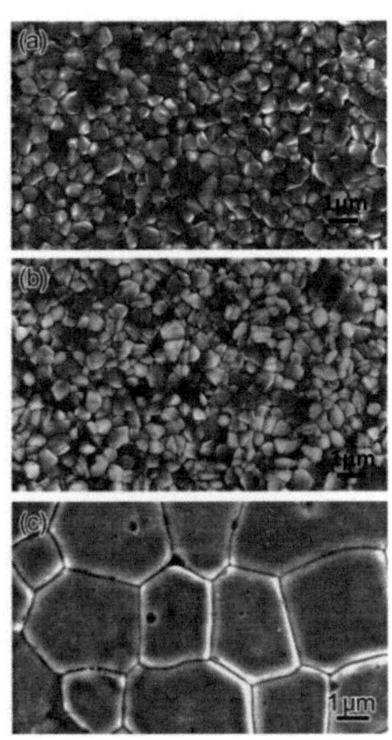

Figure 3.
The example of SEM images of 2Y, 3Y, and 8Y sintered at 1500°C [2].

ceramics material be produced. Grain size played a very important role in affecting the material properties which included hardness, fracture toughness, and elastic modulus. By sustaining a particular phase at room temperature, it allows YSZ to function specific role in the industry as specific material properties are induced. In this chapter, thermal barrier coatings (TBC), solid oxide fuel cells (SOFC) and biomaterial, which are the common industry applications, will be explained and discussed. Each application will highlight the advantages as well as the limitations respectively. All these are related with phase retention, which beyond the zirconia natural phase transformation toward the temperature.

2. Thermal barrier coating (TBC)

2.1 The working principles

Thermal barrier coating (TBC) refers to a thermal insulation layer for the purpose of protecting the original material (base material) at high service operation temperature and hot corrosion environment. The described situation normally happens in heavy duty industry like aircraft, automotive, offshore equipment (power generator). In order to protect and secure the base material from extreme scenario and further prolong the lifespan of the machine and system, TBC is one of the option to be used [3]. The common components will be applied the coating included but not limited, of diesel engine and gas turbine blade [4, 5].

Figure 4 illustrated the concept, construction, and function of thermal barrier coating system. It basically consisted of a surface layer (or been called as "top-coat (TC)") to form a bond coat (BC) layer. In between these two layers, a thin layer called Thermal Grown Oxide (TGO) is formed due to high-temperature oxidation of the bond coat [6]. The three layers function as the coating to protect the base materials from extreme service environment.

YSZ is one of the common ceramic materials to form the TC layer and the BC layer is made of different types of metallic material. This is due to YSZ possessing thermal properties like low thermal conductivity and high coefficient of thermal expansion. This main role of BC layer is to generate an adhesion with TC, then to protect the base material from oxidation [7]. Many research showed that 6–8 mol% yttria stabilizer

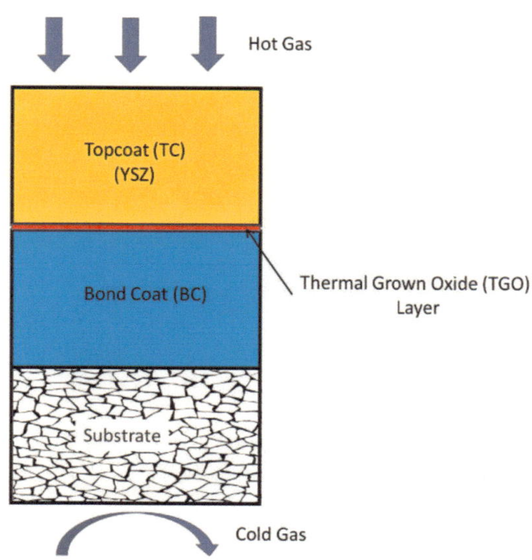

Figure 4.
Illustration of thermal barrier coating.

zirconia (6-8YSZ) exhibited excellent thermal properties for example, low thermal conductivity, and high thermal expansion coefficient [8]. The phases of this level normally consist of tetragonal and cubic, or both phases co-exist at the same time.

TBC can be constructed by two common methods—thermal spraying and electron beam-physical vapor deposited (EB-PVD). Thermal spraying methods included atmospheric-plasma spray (APS), plasma spray-physical vapor deposition (PS-PVD) and high-velocity oxy-fuel (HVOF) spraying [9]. The basic mechanism of thermal spraying is to melt the coating material and propel it into the substrate material with high and strong velocity. Plasma spraying basically created the scenario through plasma jet and HVOF is to use high-velocity jet instead.

It is more common to adopt thermal spraying method for made up of TBC for a few reasons. First, thermal spraying was widely accepted by the industry with the technique compatible with various materials, which included ceramics, metals, and alloys. Second, low operational cost and third, low equipment specification requirement. All these caused the mentioned technique received high welcoming small and medium-sized factory. However, many previous works had reported EB-PVD method did produce a higher quality result compared with plasma spray methods [10, 11]. Even though in recent years, new and advanced plasma spray techniques are emerging, the full exploration of the method is yet to be fully accepted by the industry. For example, new techniques like high-velocity air fuel (HVAF) and liquid feedstock thermal spraying, had claimed it able to produce a high quality of surface by increasing the adhesive strength of the coating [12]. The technique of HVAF spraying environment is showed in **Figure 5**.

2.2 Limitation and solutions

Even though YSZ is a proven material to form TC, degradation is still becoming one of the biggest challenges, and the major failure mechanisms included residual stress, hot corrosion, oxidation, and phase transformation [13]. All these became the factors of shortened the life span of TBC and hinder it continuing to perform in an optimum condition. The durability of TBC refers to the high quality of adhesive strength and low thermal conductivity (insulation) at high temperature [14]. Once these two capabilities been destroyed, TBC basically will fail under the tough operation and service environment, and the mentioned failure mechanisms are playing the role to degrade TBC.

Phase transformation can be related to degrading TBC through the tetragonal transformation back to monoclinic when thermal gradient occurs during the

Figure 5.
High-velocity air fuel (HVAF) technique [12].

operation. The happened due to the oxygen diffused to bond coating layer and allow chemical reaction with YSZ. This had greatly decreased the amount of oxygen of YSZ. Since oxygen ion is one of the factors to stabilize tetragonal phase by forming chemical bonding with zirconium ion, decreasing of the amount of the ion directly destabilized tetragonal phase and drove the phase transformation happen. Besides that, the cooling of TBC also contributed to the phase transformation, as that is the natural characteristic of zirconia. However, this mechanism does not occur immediately. Various research has been conducted to overcome the challenges of TBC to prolong the service life, and one of the common methods is by using dopant. Co-doping Yb_2O_3-Gd_2O_3-Y_2O_3 co-doped with ZrO_2 (YGYZ), the outcome showed the tetragonal phase able to be retained 40% compared to the undoped YSZ [15].

The X-ray diffraction (XRD) result revealed that the doped sample consisted of stable tetragonal phase greater than the undoped YSZ, which is showed in **Figure 6**. Take note that YVO_4 is the corrosion product, due to the experimental testing by using Na_2SO_4 + V_2O_5 molten salts to simulate the real hot corrosion environment. In another word, the doping technique also improved the hot corrosion resistance of YSZ. This research outcomes in **Figure 7** showed a similar trend of creating double layer of topcoats by using $La_2Zr_2O_7$ under the similar experimental condition [16].

Slow cooling rate is another finding to avoid phase transformation, as well as to prolong the life span of TBC as showed in **Figure 8**. The research reported by reducing the cooling rate to 10 K/s, compared to 100 K/S, which is the common practice, also increased the operational temperature from 1200°C to 1500°C [17]. All the research revealed that doping (or co-doping) is an effective solution to minimize the phase transformation, which will avoid the mentioned failure mechanisms.

Besides avoiding the failure mechanism, the advantages of doping technique also bring benefits and improvement to YSZ as TBC. By doping TiO_2 to YSZ, as one of the recent finding, showed the dopant able to increase the operational temperature of TBC from 1200°C to 1600°C [18]. In **Figure 9**, it also resulted the doped samples (TZ) had a lower thermal conductivity compared with the undoped sample (YSZ). The research claimed that low thermal conductivity was attributed by the lattice disorder, which restrict the movement of phono and lower the thermal conductivity.

Recent year, a method calls sol-gel method to fabricate YSZ aerogel had been developed. The research outcome reported the new coating successfully lower down

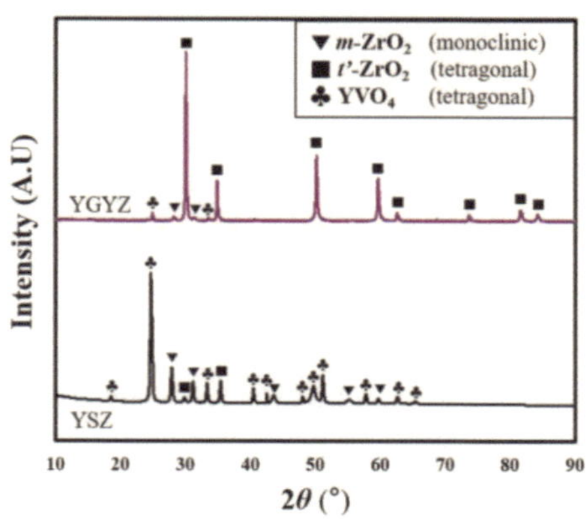

Figure 6.
XRD analysis after hot corrosion at 1100°C [15].

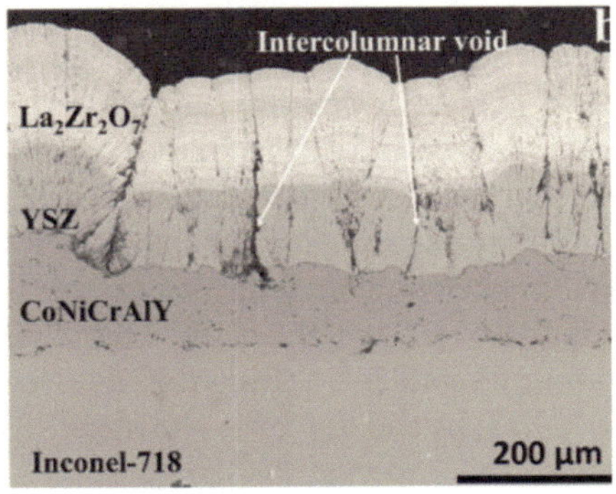

Figure 7.
SEM image of double layer topcoat [16].

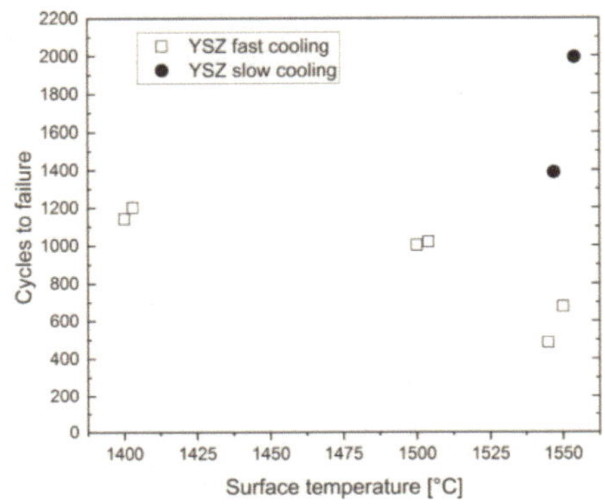

Figure 8.
Number of cycles to failure for YSZ as TBC [17].

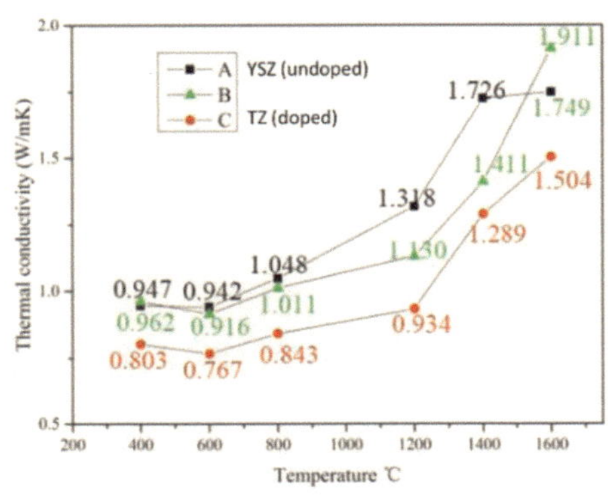

Figure 9.
Thermal conductivity of doped and undoped YSZ [18].

the thermal conductivity through a finer porosity in the aerogel structure [19]. The conceptual illustration of the method on gas turbine is illustrated in **Figure 10**.

This earlier work revealed that applying dopant to YSZ and using different sintering method are one of those effective strategies to overcome the limitation which been mentioned in TBC application. Some research work even showed that doping and sintering methods can affect the average grain size which further affected

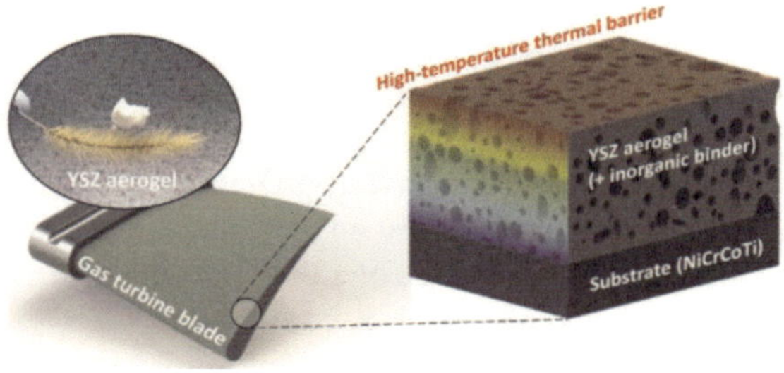

Figure 10.
Conceptual illustration of the YSZ aerogel TBC coating on the gas turbine blade [19].

the mechanical properties of the materials. However, the effect also depends very much on the substrates to be used.

3. Solid-oxide fuel cells (SOFC)

3.1 The working principles

Green technology and energy are becoming one of the major trends globally, and the Department of Economic and Social Affairs, United Nations has established 17 sustainable development goals (SDGs). Without any surprise, clean energy and a green environment are one of the major aspects.

Solid Oxide Fuels Cell (SOFC) received great attention for supply of environmental-friendly energy. One of the main advantages of SOFC is to generate power from hydrogen, natural gas, and other renewable fuels. Based on **Figure 11**, the SOFC reaction can be explained by the following chemical equation:

Anode:

$$H_2 + O^{2-} \rightarrow H_2O + 2e- \quad (1)$$

$$CO + O^{2-} \rightarrow CO_2 + 2e- \quad (2)$$

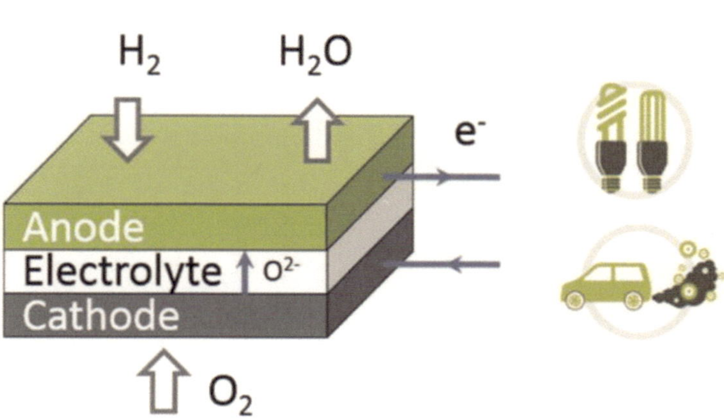

Figure 11.
Illustration diagram of SOFC [20].

Cathode:

$$O_2 + 2e- \rightarrow 2O^{2-} \quad (3)$$

By burning the fuel (hydrogen or hydrocarbon) as the input on anode, the concentration of oxygen, from the surrounding environment (air), will be "consumed," as illustrated in Eqs. (1) and (2). The electrons will be transported through external circuits for industry applications. Meanwhile, oxygen on cathode side will react with the electron been transported and produced oxygen ion, as shown in Eq. (3). The porous character of electrolyte will allow the oxygen ions to diffuse to anode and continue the cycle. The whole operation needs to be happened at high operating temperature in between 800°C and 1000°C [21]. This operation of conversion greatly reduced the gas emission (heat), which is bringing air pollution to the environment, and the other product from the reaction will be water (H_2O) [22].

YSZ is one of the promising materials to be utilized as solid electrolyte material for SOFC. A qualify material as the medium and play the role of solid electrolyte required characteristic of conducting ions effectively. 8YSZ possess characteristics of high concentration of oxygen ion vacancies, which allow an effectiveness and efficiency performance of SOFC, due to 8 mol% yttria able to retain cubic phase (c-YSZ) of zirconia, which allow highest oxygen ion vacancies concentration compare to the other phase of zirconia [23].

3.2 Limitation and solution

However, utilizing 8YSZ as the electrolyte of SOFC comes with restrictions and limitations which require the high operating temperature to "activate" SOFC to reach high ion conductivity. In such an operational environment, the components of SOFC may encounter thermal expansion and contraction and further consume the durability of the cell. To overcome this, it drives the scientist and engineers to study and develop different types of materials to widen the application of SOFC. At the same time, it also opened the possibility of continuously enhancing and improving 8YSZ without compromising the conductivity yet continues to lower the operational temperature as well as the mechanical properties [24]. 8YZP basically exhibited low mechanical properties, so several research and investigation had been conducted to reach a breakthrough of these limitations.

Instead of using conventional sintering method, flash sintering method had be proven as the sintering method not only increase the ionic conductivity of 8YZP as

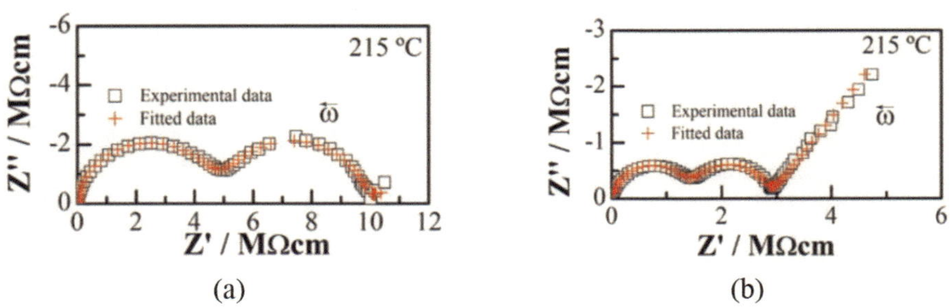

Figure 12.
Nyquist plot of 8YSZ with (a) conventional sintering (b) flash sintering method at 215°C [26].

electrolyte in SOFC, at the same time, produced 8YSZ to reach full densification at a lower sintering temperature [25].

In **Figure 12**, the Nyquist plot, which generated by Electrochemical Impedance Spectroscopy (EIS), indicated flash sintering produce higher conductivity, smaller grain size and porosity. In the plot, flash sintering showed the semicircle with smaller diameter, which referred to a smaller grain size. Smaller grain size offered higher surface area for more ion oxygen to move or transport. This factor attributed a higher conductivity of the ceramic material [26].

Besides sintering method, utilizing dopant also become another strategy to increase the ionic conductivity of 8YSZ. Dopant included iron(III) oxide (Fe_2O_3) reported an satisfactory result by increasing the ion conductivity [27].

The result in **Figure 13** was generated through molecular dynamics (MD) simulation, and it showed that the optimum result of the conductivity was 4 wt% of Fe_2O_3 dopant. The simulation result agreed with another research outcome by using the same dopant [28]. The research claimed that doping can stabilize cubic phase of 8YSZ, which is high of ion conductivity, through creating oxygen vacancies.

The result of XRD in **Figure 14** validated the claim, which showed the doped sample consist high volume of cubic phase regardless the sintering temperature. By using the cold sintering process (CSP), the study also showed the enhancement of ion conductivity compared with conventional sintering method.

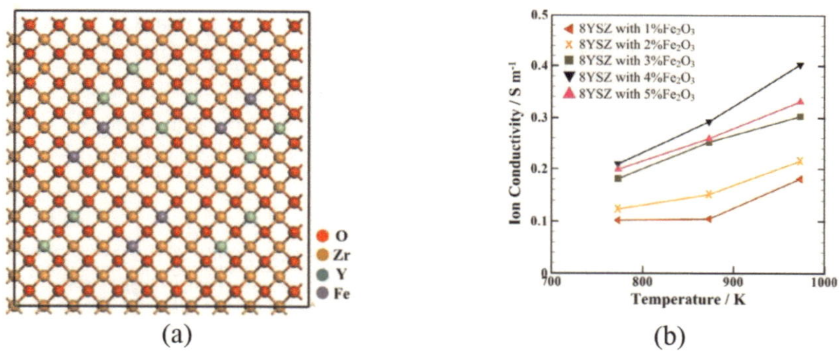

Figure 13.
8YSZ doped with Fe_2O_3 of: (a) MD simulation diagram and (b) oxygen ion conductivity of [27].

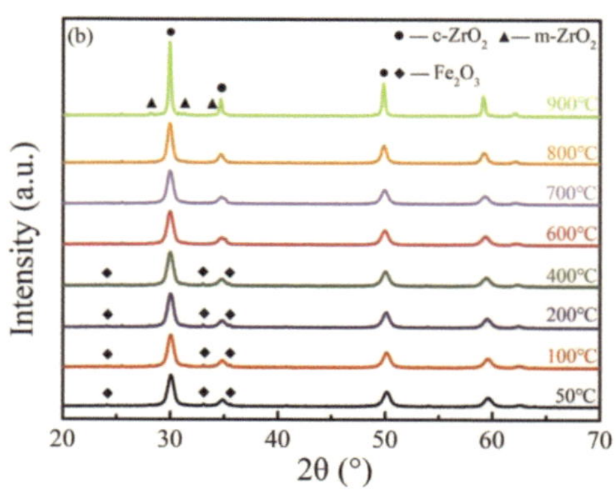

Figure 14.
XRD pattern of 4 wt% Fe_2O_3-doped 8YSZ sintered by using cold sintering process [28].

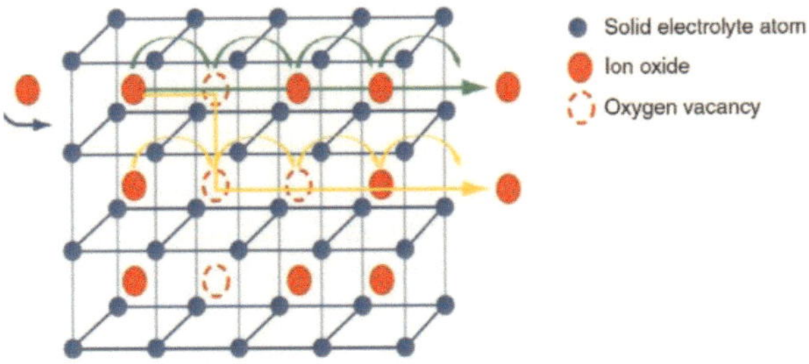

Figure 15.
The illustration diagram of ion migration mechanism [29].

Figure 15 illustrated the how doping created oxygen vacancies and allow ion to migrate and increased the ion conductivity characteristic. Dopant will introduce lattice defects and distortions, which produced oxygen vacancies in the ceramic material. The vacancies allowed migration of oxygen ion, thus increasing the ionic conductivity of the material [29].

When entering millennium year, the doping method on 8YSZ started to focus on improving mechanical properties, which is important to enhance the durability of the solid electrolyte in the cell. A study had been conducted by using 8 mol% Lu_2O_3 doped 8YSZ and successfully improved the flexural strength of the materials without compromising the ion conductivity [30]. Other studies on the dopant like CuO, TiO_2 and Bi_2O_3, also reported the sintering aids for 8YSZ effectively aided the ceramics material to reach full densification at a lower sintering temperature without compromising ion conductivity [31–33]. Lower sintering temperature for full densification, which is highly related to the mechanical properties like hardness and fracture toughness, means less energy required during processing.

Transition metal oxide doping strategy showed significantly of improving the mechanical properties of 8YSZ. However, an interesting study and investigation had been conducted. Instead of using transition metal oxide as the dopant, 3YZP (10, 25 and 35 wt.%) had been doped to 8YSZ and both mechanical and electrical properties had been evaluated. The study reported the doped material had an improved version in terms of Vickers' hardness and fracture toughness. At the same time, ionic conductivity also increased, but with only happen under the service temperature below 550°C [34].

Sintering methods and sintering aids (dopant) had been the main factor to enhance the ionic conductivity of the materials and elevate the performance of 8YSZ as the solid electrolyte in SOFC. While the mentioned investigation also revealed doping strategies able to improve the mechanical properties of the material by controlling the amount of the dopant.

4. Biomaterials

4.1 The working principles

The phase diagram (**Figure 2**) showed the amount of yttria stabilizer determine the phase of retention of zirconia (monoclinic, tetragonal, or cubic), and 3 mol% yttria tetragonal zirconia polycrystal (3Y-TZP) is one of the ceramic materials received great welcoming as biomaterial. Among all the three phases of zirconia, tetragonal phase exhibited high hardness strength and fracture toughness [35]. Besides that, zirconia also showed a friendly biocompatible characteristic, which is greatly benefit to bone and tissue implantation [36]. The high fracture toughness is attributed by a phenomenon called stress-induced transformation toughening. This mechanism happens when cracking happens on the surface of 3Y-TZP, metastable tetragonal will transform back to monoclinic, and this transformation will increase the toughness of the material, as illustrated in **Figure 16**.

Due to these attractive mechanical properties, ceramic material is usually utilized in biomaterial industry to enhance the mechanical structure of the product. Besides the mentioned material properties, 3Y-TZP also showed itself with the criteria as the bio-ceramic, which include high corrosion and wear resistance, and esthetics [37].

4.2 Limitation and solution

However, 3Y-TZP does come with limitations and the most common one is called low temperature degradation (LTD) or hydrothermal aging. It is a phenomenon when 3Y-TZP exposed to moisture or humidity environment (water or water vapor presence) at temperature between 65°C and 400°C, which commonly for biomaterial application [38]. Such condition will allow phase transformation, from tetragonal (*t*) to monoclinic (*m*), on the surface of the material then produced intergranular cracking [39]. This is because the tetragonal phase of 3Y-TZP is a metastable state at room temperature and transformed back to monoclinic phase, which is the most stable phase of zirconia at room temperature, will easily take place under such environment. Once monoclinic phase saturated and dominated the microstructure of the material, the mechanical properties will also degrade. It received a great welcome from the biomedical industry to study and investigate the factors to improve and overcome the mentioned challenge. However, the effect of LTD did not happen immediately but the degradation will take long period of time. In order to study LTD effectively, scientists and researchers simulated the condition under the environment of laboratory. The condition commonly by using conventional water-vapor autoclave operating at low temperature (98, 121 and 132°C) respectively and under adiabatic pressure for different exposure time [40]. In

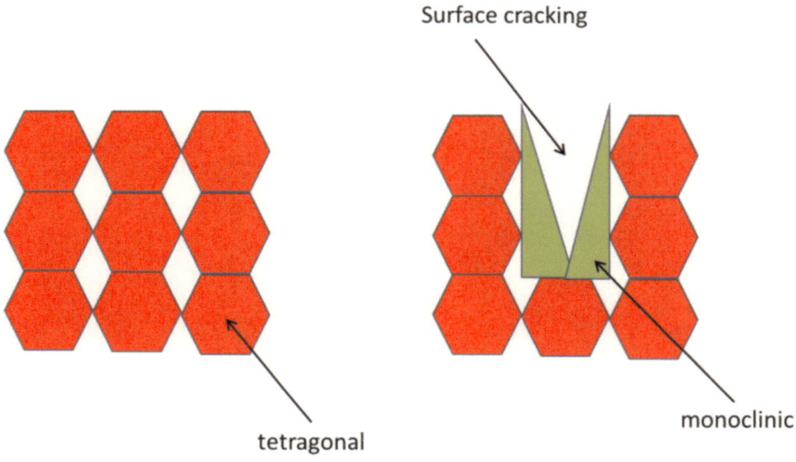

Figure 16.
Illustration of transformation toughening of 3Y-TZP.

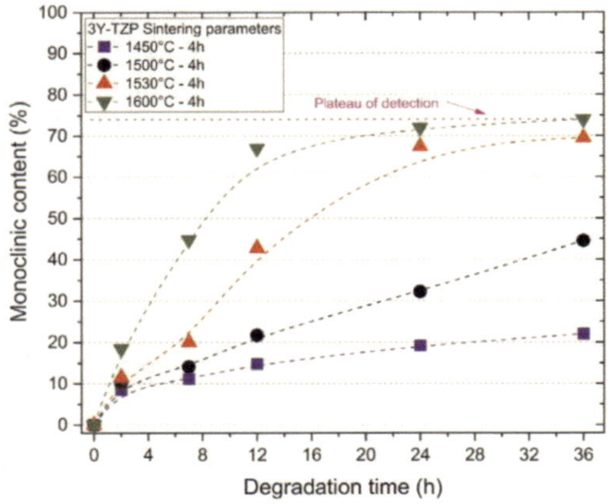

Figure 17.
Monoclinic vs degradation time of 3Y-TZP under degradation exposure time (134°C, 4 bar) [41].

Figure 17 showed the example of the evaluation of the phase transformation (t - > m) when exposed the simulated degradation environment [41]. The method has been proven as an effective way to assess any modified and improved 3Y-TZP of the characteristic toward the resistance of LTD.

Several studies showed doped 3Y-TZP exhibited good LTD resistance by delaying the phase transformation then further minimized the degradation of mechanical properties. Co-doping CaO and CeO_2 to 3Y-TZP exposed to the simulated LTD environment, the result in **Figure 18(a)**, reported that the no monoclinic phase exists in the microstructure of the samples regardless the amount of both the dopants [42].

Manganese dioxide (MnO_2) (0.5 wt% and 1.0 wt% respectively) achieved a similar result under the same LTD laboratory condition but exposed to a longer holding time, 120 h, which showed in **Figure 18(b)**. The dopant effectively increase the LTD resistance of 3Y-TZP [43]. Both XRD diagrams also revealed a stable tetragonal phase in the microstructures, which enhanced the mechanical properties like hardness and fracture toughness of the doped 3Y-TZP as well.

Besides the mentioned dopant, other transition metal oxide like Copper Oxide (CuO), graphene oxide (GO), Flaysh also had been studied and investigated and

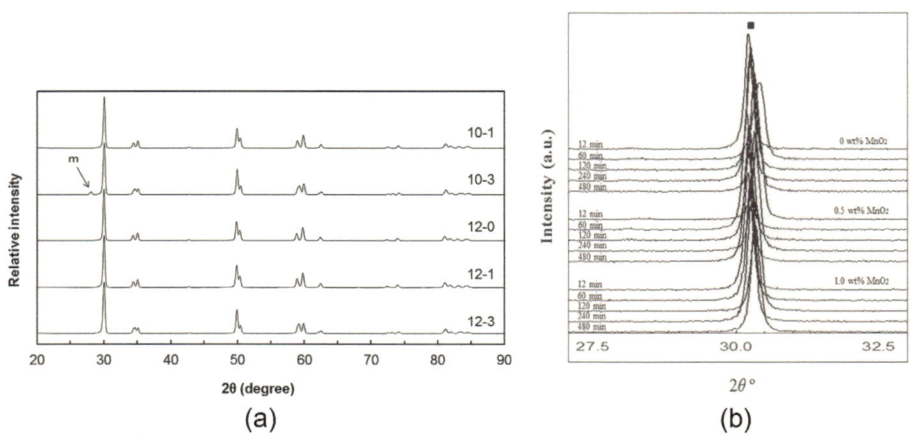

Figure 18.
XRD pattern of the samples after degradation exposure time of (a) 30 h [42] and (b) 120 h [43].

achieved a similar outcome [44–46]. Some research even brings further insight by relating the LTD resistance of 3Y-TZP with grain size. It reported grain size less than 3 μm will improve the LTD resistance and beyond the mentioned size, the effect will not take place. And doping Ceria and Alumina to 3Y-TZP able to control the grain size below the mentioned level effectively [47].

Sintering condition, which included sintering temperature and sintering holding time, also played a vital role for improving the LTD resistance of 3-YZP. The sintering temperature, 1450–1650°C and holding time, 1, 2 and 4 h, became the study scope. The outcome revealed that the higher both the mentioned parameters, the higher sensitivity of LTD of 3Y-TZP [48]. In year 2018, a research showed that the level of LTD resistance as well as the microstructure of 3Y-TZP had been improved through different sintering cycle and conditions [49]. The effect of pressureless sintering with a range of temperature, 1400–1600°C, and to produce an optimum result between the resistance of LTD, hardness and fracture toughness also had been studied. The outcomes of the study showed that the optimum sintering temperature, 1500°C, is the best temperature to balance the three (3) mentioned material properties of 3Y-TZP [50].

There is a huge potential to continue to improve 3Y-TZP application for biomedical related industries. Both sintering and doping methods have shown promise in enhancing the aging resistance and mechanical properties of 3Y-TZP, making it more suitable for use in biomedical applications. Ongoing research in these areas is expected to further enhance 3Y-TZP's potential for biomedical use and lead to new innovations. Therefore, both methods are valuable tools in unlocking the potential of 3Y-TZP for biomedical applications.

5. Conclusion

YSZ is a versatile material that can be applied in various industries, which include TBC, SOFC, and biomaterials.

In the context of TBC, YSZ has become the promising material due to its material characteristic to withstand high-temperature environments and protect underlying components from thermal damage. This makes it ideal for use in gas turbine engines and other high-temperature applications.

In the field of SOFC, YSZ is commonly used as an electrolyte due to its high ionic conductivity and stability under operating conditions. The use of YSZ electrolytes has contributed to the development of high-performance SOFCs, which have the potential to play a significant role in the future of clean energy production.

YSZ's biocompatibility and high mechanical properties like hardness and fracture toughness make it a suitable material especially for implants, and other biomedical structural-related applications.

The use of different sintering methods and various dopants played an important role in to overcome the different challenges and able to enhance the material properties of YSZ, such as its crystal structure (phase) stability, ionic conductivity, and mechanical properties. Even though it seems that thermal properties, electrical properties, and mechanical properties are emphasized in TBC, SOFC and biomaterials respectively, it can be challenging to achieve significant improvements in all properties simultaneously. Therefore, it is important to prioritize specific properties based on the requirements of each application.

Author details

Alexander Chee Hon Cheong* and SivaKumar Sivanesan
Asia Pacific University Technology and Innovation, Kuala Lumpur, Malaysia

*Address all correspondence to: alexander@apu.edu.my

References

[1] Alexander Chee HC, Singh RSK, Sara Lee KY. Effects of pressureless two-step sintering on the densification and properties of tetragonal zirconia. Journal of Ceramic Processing Research. 2021; **22**(3):289-295

[2] Matsui K, Yoshida H, Ikuhara Y. Grain-boundary structure and microstructure development mechanism in 2–8 mol% yttria-stabilized zirconia polycrystals. Acta Materialia. 2008; **56**(6):1315-1325

[3] Li D, Jiang P, Gao R, Sun F, Jin X, Fan X. Experimental and numerical investigation on the thermal and mechanical behaviours of thermal barrier coatings exposed to CMAS corrosion. Journal of Advanced Ceramics. 2021;**10**(3):551-564. DOI: 10.1007/s40145-021-0457-2

[4] Godiganur VS, Nayaka S, Kumar GN. Thermal barrier coating for diesel engine application—A review. Materials Today: Proceedings. 2021;**45**:133-137. Available from: https://www.sciencedirect.com/science/article/pii/S2214785320376902

[5] Sankar V, Ramkumar PB, Sebastian D, Joseph D, Jose J, Kurian A. Optimized thermal barrier coating for gas turbine blades. Materials Today: Proceedings. 2019;**11**:912-919. Available from: https://www.sciencedirect.com/science/article/pii/S2214785318329353

[6] Chellaganesh D, Khan MA, Jappes JTW. Thermal barrier coatings for high temperature applications—A short review. Materials Today: Proceedings. 2021;**45**(xxxx):1529-1534. DOI: 10.1016/j.matpr.2020.08.017

[7] Mondal K, Nuñez L, Downey CM, van Rooyen IJ. Recent advances in the thermal barrier coatings for extreme environments. Materials Science for Energy Technologies. 2021;**4**:208-210. Available from: https://www.sciencedirect.com/science/article/pii/S2589299121000215

[8] Vaßen R, Bakan E, Mack D, Schwartz-Lückge S, Sebold D, Jung Sohn Y, et al. Performance of YSZ and Gd2Zr2O7/YSZ double layer thermal barrier coatings in burner rig tests. Journal of the European Ceramic Society. 2020;**40**(2):480-490. Available from: https://www.sciencedirect.com/science/article/pii/S0955221919306892

[9] Thibblin A, Jonsson S, Olofsson U. Influence of microstructure on thermal cycling lifetime and thermal insulation properties of yttria-stabilized zirconia thermal barrier coatings for diesel engine applications. Surface and Coating Technology. 2018;**350**(July):1-11. DOI: 10.1016/j.surfcoat.2018.07.005

[10] Xia J, Yang L, Wu RT, Zhou YC, Zhang L, Huo KL, et al. Degradation mechanisms of air plasma sprayed free-standing yttria-stabilized zirconia thermal barrier coatings exposed to volcanic ash. Applied Surface Science. 2019;**481**:860-871. Available from: https://www.sciencedirect.com/science/article/pii/S0169433219307044

[11] Boissonnet G, Chalk C, Nicholls JR, Bonnet G, Pedraza F. Phase stability and thermal insulation of YSZ and erbia-yttria co-doped zirconia EB-PVD thermal barrier coating systems. Surface and Coating Technology. 2020;**389**: 125566. Available from: https://www.sciencedirect.com/science/article/pii/S0257897220302358

[12] Joshi S, Nylen P. Advanced coatings by thermal spray processes.

Technologies. 2019;7(4):79. Available from: https://www.mdpi.com/2227-7080/7/4/79

[13] Mehta A, Vasudev H, Singh S. Recent developments in the designing of deposition of thermal barrier coatings—A review. Materials Today: Proceedings. 2019;**26**(xxxx):1336-1342. DOI: 10.1016/j.matpr.2020.02.271

[14] Mehboob G, Liu M-J, Xu T, Hussain S, Mehboob G, Tahir A. A review on failure mechanism of thermal barrier coatings and strategies to extend their lifetime. Ceramics International. 2020;**46**(7):8497-8521. Available from: https://www.sciencedirect.com/science/article/pii/S0272884219337022

[15] Song D, Song T, Paik U, Lyu G, Kim J, Yang S, et al. Hot-corrosion resistance and phase stability of Yb2O3–Gd2O3–Y2O3 costabilized zirconia-based thermal barrier coatings against Na2SO4 + V2O5 molten salts. Surface and Coating Technology. 2020;**400**: 126197. Available from: https://www.sciencedirect.com/science/article/pii/S0257897220308665

[16] Ozgurluk Y, Doleker KM, Ahlatci H, Karaoglanli AC. Investigation of calcium–magnesium-alumino-silicate (CMAS) resistance and hot corrosion behavior of YSZ and La2Zr2O7/YSZ thermal barrier coatings (TBCs) produced with CGDS method. Surface and Coating Technology. 2021;**411** (January):126969. DOI: 10.1016/j.surfcoat.2021.126969

[17] Vaßen R, Mack DE, Tandler M, Sohn YJ, Sebold D, Guillon O. Unique performance of thermal barrier coatings made of yttria-stabilized zirconia at extreme temperatures (>1500°C). Journal of the American Ceramic Society. 2021;**104**(1):463-471. DOI: 10.1111/jace.17452

[18] Yang M, Zhu Y, Wang X, Wang Q, Ai L, Zhao L, et al. Preparation and thermophysical properties of Ti4+ doped zirconia matrix thermal barrier coatings. Journal of Alloys and Compounds. 2019; **777**:646-654. Available from: https://www.sciencedirect.com/science/article/pii/S0925838818341513

[19] Yoon S, Han GD, Jang DY, Kim JW, Kim DH, Shim JH. Fabrication of yttria-stabilized zirconia aerogel for high-performance thermal barrier coating. Journal of Alloys and Compounds. 2019; **806**:1430-1434. DOI: 10.1016/j.jallcom.2019.07.156

[20] Hao SJ, Wang C, Le LT, Mao Z-M, Mao Z-Q, Wang JL. Fabrication of nanoscale yttria stabilized zirconia for solid oxide fuel cell. International Journal of Hydrogen Energy. 2017;**42**(50): 29949-29959. DOI: 10.1016/j.ijhydene.2017.08.143

[21] Zakaria Z, Abu Hassan SH, Shaari N, Yahaya AZ, Boon KY. A review on recent status and challenges of yttria stabilized zirconia modification to lowering the temperature of solid oxide fuel cells operation. International Journal of Energy Research. 2020;**44**(2):631-650. Available from: https://onlinelibrary.wiley.com/doi/10.1002/er.4944

[22] Zakaria Z, Awang Mat Z, Abu Hassan SH, Boon KY. A review of solid oxide fuel cell component fabrication methods toward lowering temperature. International Journal of Energy Research. 2020;**44**(2):594-611. Available from: https://onlinelibrary.wiley.com/doi/10.1002/er.4907

[23] Jaiswal N, Tanwar K, Suman R, Kumar D, Uppadhya S, Parkash O. A brief review on ceria based solid electrolytes for solid oxide fuel cells. Journal of Alloys and Compounds. 2019;

781:984-1005. DOI: 10.1016/j.jallcom.2018.12.015

[24] Dwivedi S. Solid oxide fuel cell: Materials for anode, cathode and electrolyte. International Journal of Hydrogen Energy. 2020;**45**(44): 23988-24013. DOI: 10.1016/j.ijhydene.2019.11.234

[25] Zhang J, Wang Z, Jiang T, Xie L, Sui C, Ren R, et al. Densification of 8 mol % yttria-stabilized zirconia at low temperature by flash sintering technique for solid oxide fuel cells. Ceramics International. 2017;**43**(16):14037-14043. DOI: 10.1016/j.ceramint.2017.07.137

[26] Vendrell X, Yadav D, Raj R, West AR. Influence of flash sintering on the ionic conductivity of 8 mol% yttria stabilized zirconia. Journal of the European Ceramic Society. 2019;**39**(4): 1352-1358. Available from: https://linkinghub.elsevier.com/retrieve/pii/S0955221918307672

[27] Chan YH, Lai HY, Chen CK. Enhancing oxygen iron conductivity of 8YSZ electrolytes in SOFC by doping with Fe2O3. Computational Materials Science. 2018;**147**:1-6. DOI: 10.1016/j.commatsci.2018.01.057

[28] Cong L, Hanying L, Yuming H, Shiyong Z, Xiao W, Tengfei L, et al. Phase evolution and enhanced sinterability of cold sintered Fe2O3-doped 8YSZ. Ceramics International. 2020;**46**(9):14217-14223. DOI: 10.1016/j.ceramint.2020.02.230

[29] Zakaria Z, Kamarudin SK. Advanced modification of scandia-stabilized zirconia electrolytes for solid oxide fuel cells application—A review. International Journal of Energy Research. 2021;**45**(4):4871-4887. Available from: https://onlinelibrary.wiley.com/doi/10.1002/er.6206

[30] Park M, Jo K, Lee H, Yoon B, Lee H. Oxygen-ion conductivity and mechanical properties of Lu2O3-doped ZrO2 as a solid electrolyte. Ceramics International. 2021;**47**(15):20844-20849. Available from: https://www.sciencedirect.com/science/article/pii/S0272884221008452

[31] Xing Y-Z, Men Y-N, Feng X, Geng J-H, Zou Z-R, Chen F-H. Evolutions in the microstructure and ionic conductivity of CuO-doped yttria-stabilized zirconia. Journal of Solid State Chemistry. 2022;**315**:123497. Available from: https://www.sciencedirect.com/science/article/pii/S0022459622006223

[32] Chasta G, Dhaka MS. TiO2 incorporated YSZ thin films for SOFCs: Thermal evolution to the microstructural, topographical and electrochemical properties. Surface and Coating Technology. 2023:129318. Available from: https://www.sciencedirect.com/science/article/pii/S0257897223000932

[33] Tian Y, Liu S, Zhang X, Xiao S, Sun J, Zhang J, et al. Controlled synthesis of Bi2O3–YSZ composite powders and their sintering behavior for high-performance electrolytes. International Journal of Applied Ceramic Technology. 2022. DOI: 10.1111/ijac.14270

[34] Ghatee M, Shariat MH, Irvine JTS. Investigation of electrical and mechanical properties of 3YSZ/8YSZ composite electrolytes. Solid State Ionics. 2009;**180**(1):57-62. DOI: 10.1016/j.ssi.2008.10.006

[35] Zhang Y, Lawn BR. Novel zirconia materials in dentistry. Journal of Dental Research. 2018;**97**(2):140-147. Available from: http://journals.sagepub.com/doi/10.1177/0022034517737483

[36] Jin KCW, Dickson K, Singh R. Mechanical characterization of zirconia

ceramic composite. In: Narayana Namasivayam S, Hosseini Fouladi M, Eunice Phang SW, Chua BL, Chow YH, Yong LC, et al., editors. MATEC Web Conf [Internet]. Vol. 152. 2018. p. 02006. Available from: https://www.matec-conferences.org/10.1051/matecconf/201815202006

[37] Zhang F, Reveron H, Spies BC, Van Meerbeek B, Chevalier J. Trade-off between fracture resistance and translucency of zirconia and lithium-disilicate glass ceramics for monolithic restorations. Acta Biomaterialia. 2019;**91**: 24-34. DOI: 10.1016/j.actbio.2019.04.043

[38] Swab JJ. Low temperature degradation of Y-TZP materials. Journal of Materials Science. 1991;**26**(24): 6706-6714

[39] Ramesh S, Sara Lee KY, Tan CY. A review on the hydrothermal ageing behaviour of Y-TZP ceramics. Ceramics International. 2018;**44**(17):20620-20634. DOI: 10.1016/j.ceramint.2018.08.216

[40] Zhu W, Fujiwara A, Nishiike N, Marin E, Sugano N, Pezzotti G. Transition metals increase hydrothermal stability of yttria-tetragonal zirconia polycrystals (3Y-TZP). Journal of the European Ceramic Society. 2018;**38**(10): 3573-3577. DOI: 10.1016/j.jeurceramsoc.2018.03.042

[41] Abreu LG, Quintino MN, Alves MFRP, Habibe CH, Ramos AS, Santos C. Influence of the microstructure on the life prediction of hydrothermal degraded 3Y-TZP bioceramics. Journal of Materials Research and Technology. 2020;**9**(5):10830-10840. DOI: 10.1016/j.jmrt.2020.07.059

[42] Turon-Vinas M, Zhang F, Vleugels J, Anglada M. Effect of calcia co-doping on ceria-stabilized zirconia. Journal of the European Ceramic Society. 2018;**38**(6): 2621-2631. DOI: 10.1016/j.jeurceramsoc.2017.12.053

[43] Ting CH, Ramesh S, Tan CY, Zainal Abidin NI, Teng WD, Urries I, et al. Low-temperature sintering and prolonged holding time on the densification and properties of zirconia ceramic. Journal of Ceramic Processing Research. 2017;**18**(8):569-574

[44] Alexander Chee HC, Singh RSK, Sara Lee KY. The effect of transition metal oxide doping on the sintering behaviour of yttria-stabilised tetragonal zirconia. Journal of Ceramic Processing Research. 2020;**21**(4):495-500

[45] Ramesh S, Khan MM, Alexander Chee HC, Wong YH, Ganesan P, Kutty MG, et al. Sintering behaviour and properties of graphene oxide-doped Y-TZP ceramics. Ceramics International. 2016;**42**(15):17620-17625. DOI: 10.1016/j.ceramint.2016.08.077

[46] Ramesh S, Chew WJK, Chee HCA, Tan CY. Sintering behaviour and properties of flyash-doped zirconia. Materials Science Forum. 2017;**894**: 85-88. Available from: https://www.scientific.net/MSF.894.85

[47] Hallmann L, Ulmer P, Reusser E, Louvel M, Hämmerle CHFF. Effect of dopants and sintering temperature on microstructure and low temperature degradation of dental Y-TZP-zirconia. Journal of the European Ceramic Society. 2012;**32**(16):4091-4104. Available from: http://www.sciencedirect.com/science/article/pii/S0955221912004281

[48] Inokoshi M, Zhang F, De Munck J, Minakuchi S, Naert I, Vleugels J, et al. Influence of sintering conditions on low-temperature degradation of dental zirconia. Dental Materials. 2014;**30**(6):

669-678. DOI: 10.1016/j.dental.2014.03.005

[49] Wei C, Gremillard L. Towards the prediction of hydrothermal ageing of 3Y-TZP bioceramics from processing parameters. Acta Materialia. 2018;**144**: 245-256. DOI: 10.1016/j.actamat.2017.10.061

[50] Amat NF, Muchtar A, Amril MS, Ghazali MJ, Yahaya N. Effect of sintering temperature on the aging resistance and mechanical properties of monolithic zirconia. Journal of Materials Research and Technology. 2019;**8**(1):1092-1101. DOI: 10.1016/j.jmrt.2018.07.017